ONE TO NINE

ONE TO NINE

The Inner Life of Numbers

ANDREW HODGES

W. W. NORTON & COMPANY
NEW YORK   LONDON

For information about permission to reproduce selections from this book,
write to Permissions, W. W. Norton & Company, Inc.,
500 Fifth Avenue, New York, NY 10110

For information about special discounts for bulk purchases, please contact
W. W. Norton Special Sales at specialsales@wwnorton.com or 800-233-4830

Manufacturing by RR Donnelley, Bloomsburg
Production manager: Devon Zahn

Library of Congress Cataloging-in-Publication Data

Hodges, Andrew.
One to nine : the inner life of numbers /
Andrew Hodges. — 1st American ed.
p. cm.
Includes index.
ISBN 978-0-393-06641-8 (hardcover)
1. Counting. 2. Numeration. I. Title.
QA113.H635 2008
513.2'11—dc22

2007047703

W. W. Norton & Company, Inc.
500 Fifth Avenue, New York, N.Y. 10110
www.wwnorton.com

W. W. Norton & Company Ltd.
Castle House, 75/76 Wells Street, London W1T 3QT

1 2 3 4 5 6 7 8 9 0

Contents

Notes, links, updates and
answers to problems are on the Web at
www.cryptographic.co.uk/onetonine

1

The Unloved One

It is a truth universally acknowledged that a single man in possession of a good fortune must be in want of a wife. So runs a famous first sentence, full of statements about One. It claims a universal truth, but there is an ironic touch in the grandeur of Jane Austen's opening, bound as it so obviously is by human pride and prejudice, and specific to the English bourgeoisie of the early nineteenth century. Even so, a modern reader finds it a good story-starter. It is well wired into human brains, tastefully pressing the buttons labelled 'sex' and 'money' in unison.

A chapter about the number One itself cannot count on any such welcome mat. Numbers do not make light conversation, and are gatecrashers at the great party of the human arts. They seem devoid of that sympathetic wit and irony which smooths the passage of letters and life. Indeed, the numbers can hardly join in the human party at all, being abstractions unable to shake hands or flirt. Where did Page One go to? – this is already Page Two, and Page One exists only by implication, possessing a One-ness only in one's mind's eye.

Even more annoying to the party-goers, is that the numbers adopt unappetising manners whenever they get the upper hand, dictating bossy, tedious and incomprehensible regulations. This is a book about the inner life of these unwelcome guests. It is, more or less, the tale as told by the mathematicians who get to know them. I hesitate to generalise because mathematicians are an un-unionised bunch, with the solidarity of a plate of spaghetti. But the assumption implicit in mathematics is one of universal truths deeper than any Jane Austen would have countenanced as an opening gambit in polite conversation.

Numbers, to mathematicians, are the same for Cleopatra or Moctezuma, for the Neanderthals and the dinosaurs, true in any galaxy, past or future. They tell stories which could be shared with extraterrestrial intelligences. When Dante wanted something for the happy inhabitants of the Paradiso to do with eternal life, he took from Platonic tradition the contemplation of mathematical truth. Not everyone would welcome this connection: the British mathematician G. H. Hardy, who as an atheist wouldn't be seen dead in the One to Nine of Dante's heaven, made a famous declaration of mathematical idealism which had nothing to do with religion.

Prime time

Hardy expressed this other-worldliness by using the concept of a *prime number*. Any number *n* can be represented as a multiplication of 1 x *n*. Put another way, 1 and *n* are always divisors of *n*. The prime numbers are those numbers which have *only* those two divisors, and so enjoy a special relationship with One. The numbers 2, 3, 5, 7 and 11, for instance, are prime. The number 6 is not. The distinction can be seen as a picture. Six objects can be arranged in a rectangle, but seven must lie on a line.

Even the United States is powerless to change the régime of the primes, and as the stars have swelled from 13 to 50, ingenious means have been necessary to deal with this fundamental problem. In 1940, as those then neatly 6 x 8 states were about to face their transformation into a superpower, Hardy published a short book, *A Mathematician's Apology*, with a statement of superhuman Number: '317 is a prime, not because we think so, or because our minds are shaped in one way rather than another, but *because it is so...*'

What view was Hardy arguing *against*? His scorn was reserved for the profilic British polymath biologist Lancelot Hogben, author in 1936 of a major popular book, *Mathematics for the Million*. Holding his nose, Hardy quoted Hogben's characterisation of mathematics as 'the grammar of size and order', needed for the planning of 'the rational society'. Hogben was expressing the Marxism of the 1930s, which saw all cultural manifestations as the outcomes of relations of production, and thus mathematics as based in human practice. Hardy was no conservative élitist, and in fact had rather radical views. He recognised that mathematics might be useful, but insisted that its utility did not explain why mathematicians did it.

Mathematicians have largely maintained a discreet radio silence, and Hardy was unusual for striking a dissonant note in public. Someone who knew him, C. P. Snow, wrote that Hardy 'didn't give a damn'. Snow himself became famous in the 1960s for trying to combine Jane Austen's legacy with insight into twentieth-century science. It is doubtful whether Snow had much success with his unifying vision, and his novels are neglected now. But his expression 'Two Cultures' has stuck as an awkward reminder of the fact that those who run the human world and dominate its discourse are generally unaware of its base in the physical world, and of the mathematics that is the only language for expressing that world. Snow used the Second Law of Thermodynamics as his example of what ought to be well known, but he might as well have used the First Law, or indeed the Zeroth Law, as his examples: it is as true now as it was in the 1960s that none of these can be referred to in polite society.

There is now an extra reason for presenting an unwelcome message about the inescapable properties of numbers, summarised in Al Gore's expression of 'inconvenient truth'. Although the question of recent, anthropogenic climate change involves many physical sciences – including all those laws of thermodynamics – the power of mathematics to predict is at the heart of the developing theory. The calculation of climate change certainly falls into Hogben's realm of the relations of human production, but the phrase 'inconvenient truth' could have come from the inconvenient Hardy himself. The Earth's atmosphere lacks the simplicity of 317: there are even more factors involved in climate science than in cosmology, where notable corrections have been made, and there are bound to be revisions to current models. And yet those models *are* essentially mathematical. Environmental campaigners who want to disrupt airports say that 'the science tells us' in terms that they feel put their case above the law, words that recall Hardy's 'it is so'. Whether one takes Hogben's viewpoint or Hardy's, the question of the potential of mathematics is newly vital.

One and I

In the 1990s, I wrote a weekly column on mathematical topics for the London newspaper, the *Observer*. The editor, Rebecca Nicolson, contacted me again in 2005 and suggested a short book with the title *One to Ten*. I saw a point of departure. For this had been done before, in a classic book by Constance Reid, *From Zero to Infinity* (1956), still in print 50 years later. Her ten chapters were headed by the numbers from Zero to Nine, each used to spark off an elegant exposition of some substantial feature of the theory of numbers. Anything I wrote was bound to resemble her plan.

I remembered it well because I learnt so much from it, perhaps most of all because through Constance Reid I had absorbed Hardy's view of other-worldly mathematics, communicated with all his charm but without his awkward-squad agenda. Her succinct and lucid pages now convey a 1950s America, fortress of world culture, but not much given to airing controversies. It is compelling in its commitment to the fascination of numbers for their own sake. Questions of motivation or usefulness hardly arise in her account. These, however, poured from another book of my childhood: this was *Man Must Measure* (1955) by none other than Lancelot Hogben. In the 1950s he blazed a new trail, using extensive graphics and photographs to pour out an encyclopedic knowledge. In his enthusiastic account, mathematical advance was as anthropogenic as carbon dioxide, the fruit of human industry, interwoven with every kind of human skill and argument. So between them, these books exposed me to two different points of view, echoing the 1930s political conflict. Fifty years later, I find myself having to re-echo them.

Given the publisher's suggestion of *One to Ten*, and the classic example of Reid's sequence from zero to nine, my decision was for *One to Nine*. The clinching factor was, of course, the worldwide popularity of the Sudoku square. In case you've been away from the solar system for a while, here is an example of the basic 3 x 3 Sudoku puzzle. The grid must be completed so that every row, column and subsquare contains just the numbers One to Nine.

			4				6	
8					9			
		3				2		
		7						4
9	1				2			
				5		3		
2				1	8			
	6						7	
			3					9

A TOUGH problem, with no obvious first step. Hint: there is only one place where the 3 can go in the central square.

The numbers entered this book saying that they made no light conversation. But if numbers *did* make light conversation, Sudoku shows how they would talk. What's more, it shows they can tell a joke. It is this: although *The Times* and other newspapers insist that the puzzle 'needs no maths', the process of solving it is an elegant miniature version of the experience of real, adult, mathematics. What *The Times* means is that it needs no school maths, reflecting the legacy of fear and anxiety generated by schools, which leaves most of their victims with a lifetime of mumbling apologetically about 'my worst subject'.

Sudoku problems are Hardy-ish in having no use whatsoever, apart from helping to keep Alzheimer's at bay. They do nothing for the once-planned society of the future, nor for the market economy which has replaced it. In fact they must have subtracted millions of person-hours from the duties of profit-making. They have none of the cosy linguistic clubbiness of cryptic crosswords. The problems need up to an hour of concentrated thought. Yet the demand for them appears insatiable.

In contrast, school mathematics teaching seems to be in a particular state of crisis. The *Guardian*, the leading British newspaper for the education business, describes it as 'needing a makeover, to make it sexy again'. Mathematicians, its writer explains, 'are bald, overweight with beards and glasses and eternally single, leading little or no social life'. Jane Austen's efforts would clearly have been doomed had she opened with such a character, and curiously, the writer for this progressive newspaper – long a leading voice for feminism – tacitly and unrealistically assumes mathematicians to be male. Another *Guardian Education* feature helpfully describes all science as 'boring, prohibitively hard, too abstract and too male, in a spoddy, won't-get-a-girlfriend kind of way'. These artless articles reveal an unlovely truth: that mass, compulsory, school maths has enjoyed about the same success as the War on Drugs.

In stark contrast to these effusions, Sudoku gains cool cus-tomers without any need for sexing up, and the *Times* cham-pion is, as it happens, a young woman. The reason, perhaps, is that it encapsulates some of the most fascinating elements of adult mathematics: elegant geometry and pure logic. You can try to solve a puzzle by guessing – but pencilling in trial solutions and rubbing them out again is likely to be a losing battle. Faith, hope, fantasy and bluster are prominent in the planning of world domination, but are powerless to solve the Sudoku square – or mathematical problems.

Just say Nought

Although we are following Sudoku with One to Nine, we cannot ignore the Zero with which Constance Reid chose to begin. There is a difficulty about her choice. She says that Zero, 'first of the numbers, was the last to be discovered.' This makes One her *second* number, and Two her third. This seems very odd, but it makes sense if you distinguish *cardinal* and *ordinal* numbers. Word roots show how human culture has thought of 'one' in more than one way. The root *one, un, uno, ein...* answers the question of 'how many'. It is the *cardinal number* which counts. But the root that shows in *prime, prince, Fürst, first,* evoking the social relationships of primitive primates, with their prima donnas and prime ministers, is the *ordinal number*. It is the difference between 'one page' and 'Page One'. So the apparent contradiction can be explained; Constance Reid took the cardinal Zero to correspond to the ordinal 'first'.

She could have avoided this by describing Zero as the *noughth* number. The words 'noughth' or 'zeroth' do have a shadowy existence in language, used for something that has to be placed before something that has already been called 'first'. After the First Law of Thermodynamics (the conservation of energy) had been established, the Zeroth Law (temperature exists) was recognised as more primitive. Oxford University has a Noughth Week before the first week of term, but no Noughth Class degrees better than a First. There are other grey areas where language calls for starting a list with zero: the Dewey decimal classification, the Noughties for this first decade. However, having to explain the ordinal 'noughth' probably causes more problems than it solves.

With cardinal numbers, at least there is no doubt about what Zero means, and Constance Reid argued that it (probably) answers the question, 'How many elephants are there in the room where you are reading this book?' Nevertheless, we do not normally start counting until there is something to count. A queue of zero people is not a queue at all. For this reason, the numbers 1, 2, 3, 4... are usually called the *natural* numbers by mathematicians. *One banana, two bananas, three bananas, four...* shows the naturalness of the definition. *Yes we have no bananas,* although comprehensible, shows the oddness of offering zero bananas for sale.

Nought-words go far back in Indo-European languages, with thousands of years of clicking tongue against palate to deny, negate or undo. We show no reluctance to make a positive assertion of nothing. Yet counting zero is not like counting one or two. What is or are nought bananas? Nothing at all? If so, then 'We have no bananas' would mean the same as 'We have no sour grapes' – which it does not. The difficulty about asserting nought of something is that it seems to assert a potential existence of the very thing of which there are none. This is not problematic for greengrocers – but it is for atheists. 'I have zero pet unicorns' is a true statement, but it seems to assert the possibility of a very untrue one.

Ambivalence and awkwardness about Nought pervades English grammar. Sticklers for accuracy insist that Messrs. Blair and Bush should say that 'none of the terrorist threats is caused by the occupation of Iraq' because 'none' is short for 'not one' and therefore takes a singular verb. Yet Nought takes a plural. Join MySpace and read that you have 0 friends. (Or read the *Guardian* report that 'poor old mathematicians' have no friends except other mathematicians.) When giving her illustration, Constance Reid was unaware that 50 years later computers would announce 'there are 0 people in this chat-room', in conspicuously unnatural language.

Zero is not natural language. *How many E man? 0 please! Got any change? Sure, here's 0p! Media intern: salary $0. Yo emperor kewl kit:-) Yo m8 got 0 on.* Supermodels inspire Susie Orbach to say that 'size zero means you don't exist!' This plays exactly on the puzzle of Nought, the puzzle of observing something that is not there, and perhaps this puzzle explains why it took so long for zero to acquire a symbol, and so was 'the last to be discovered'.

The odd thing is that in the world of Greek mathematics, an abacus or its equivalent was used to do calculations. On an abacus, an empty wire or box plays the role of a zero. Yet the small step to a *written symbol*, corresponding to that empty space, was apparently never made. The innovation arose in India; a zero was used there by the ninth century CE, and used in just the same way as it is used today. The number 2007 means *two* thousands, *zero* hundreds, *zero* tens and seven. The value to be attached to the 2 and the 7 depends on their position: this is *place-notation* and it makes a zero essential.

The innovation was attested by the treatise of the leading Persian mathematician al-Khwarizmi: *On the Calculation with Hindu Numerals,* written in about 825 CE. A decimal inscription of 870 at the temple in Gwalior, central India, is the earliest that now survives. Zeroes spread into common Arabic use. Gerbert, who became pope in 999, explained the Arabic advance in a Latin treatise. There was no equivalent of the dot-com boom in that millennial year, however; for it was not until 1202 that Leonardo of Pisa, usually known as Fibonacci, successfully urged it in his *Liber Abacus* for practical use in backward Europe. Slowly, thereafter, the Arabic (or perhaps more accurately, Eurabian) numerals became used in commerce. One argument advanced against them was that place-notation makes it easy to defraud by adding on a nought, and even now cheques must be written in words, a long-term legacy of the suspicion of zero.

In the tenth century, Arab mathematicians had developed place-notation into expressing fractions, which now seem naturally bound up with zeroes and the decimal point. Yet the Babylonians had much earlier managed fractions in their system, even evaluating the square root of two to a remarkable accuracy, without a zero. It is hard to see how they could have developed one abstraction without the other. It is certainly hard now to imagine how feats such as the recently discovered classical astrolabe, gear wheels imitating accurately the motion of the Moon, could have been carried out without writing down zeroes – if only to note down the state of an abacus in mid-calculation.

Perhaps the abacus workers of the ancient world did sometimes jot down dots and dashes, but were deterred by grander literary personages from using such sad and spoddy scribbles in official writing. The problem of the calendar also gives an example of how close you can come to a concept of zero without actually getting there. The sixth-century Byzantine monk Dionysius Exiguus famously defined the Christian era with an AD1. But his method for defining the date of Easter calls for writing down the remainder when the year is divided by 19. When there is no remainder, this cries out for a zero symbol, to be regarded as on a par with the symbols for other numbers. But the Hellenic world still didn't get it. In contrast, Hindu calendars happily began with a Year Zero.

Was the Indian zero discovered or invented? Was it there all the time, as in Hardy's world of absolute truth – or was it an example of what Hogben called 'Mathematics in the Making'? A better word perhaps is *realised,* with its double meaning: both seeing a truth, and making it physically concrete.

Although I have made a fuss about the puzzle of zero being called the 'first' number, I should not leave the impression that this is a major issue. It is only for counting and ordering the *infinite* that the question of cardinals and ordinals acquires serious substance. For bananas and for book chapters, these are essentially questions of names and conventions, and a much more important point is that mathematics is a world where names and conventions do not matter, as long as they are clear. As an example, *time* is often described as a *fourth* dimension, coming after the three dimensions of space. But it is the usual convention now to place time *before* space as the *noughth* dimension.

A rose by any name

This brings us to an important distinction, on which

Constance Reid was wonderfully clear. The *numerals*, the symbols 0 1 2 3 4 5 6 7 8 9, are not numbers, but the names of numbers. Originally, telephones had dials physically sending ten clicks for a 0, and so made it made sense to put the symbols in the order 1234567890. (You cannot send zero clicks, because sending zero clicks can't be distinguished from not-sending: this is exactly the unnaturalness of nought.) Nowadays, the 0 on a telephone is just an arbitrary symbol, and telephone numbers could be as freely assigned as internet names.

The properties of numbers don't depend on the numerals used to describe them. There are many ways in which they can be described. The simplest way to write the natural numbers is by the *unary* notation of a prisoner scratching IIIIIIIIIII... on a cell wall. This shows the problem of unnatural Zero again: you cannot scratch nought marks on a prison wall to show you have not been inside.

Roman numerals can be thought of as an extension of unary notation, with a shorthand for I's in handfuls of five: V for IIIII, X for IIIII IIIII and then going on to fifties, hundreds, five-hundreds and thousands. (Thus the famous number 666, which is 500 + 100 + 50 + 10 + 5 +1, is represented as DCLXVI.) Roman usage shortened IIII to IV, VIIII to IX, a striking fact because it introduced the idea that when placed *before* a V or X, an I serves an an instruction to *subtract* one, not to add one; likewise for X and C. Thus, film copyrights dated MCMXCIX (but not MIM) preceded MM. Having conceded that the meaning of symbols may change according to their placement, it is hard to see why it took so long for the Indian place-notation to be accepted in Europe. It seems that it was the symbol 0 that was the critical factor.

But using a *zero* does not in any way oblige a counting in *tens*. The logic of place-notation works just as well if some other number is used as a base. It is not obvious why all human cultures seem to have marked off numbers in tens. The exceptions, as in the sixties of the Babylonians, still build on tens. You have to look hard to find the Amazonian Pirahã people, said to live without abstraction or numbers at all. The universality of ten is not fully explained by the form of human hands. It would surely have been just as natural to use the eight fingers, not including thumbs. Eight-based counting would have been better suited to the repeated splitting of differences which is so natural for sharing and trading, as indeed for music.

If the use of the *eight* fingers had been adopted for counting, then only the numerals 1234567 would be needed in addition to the 0. Eight would be written as 10, and what we call sixty-four would be 100. In this *octal* system, adding a 0 means multiplying by eight. The multiplication table is mercifully reduced to:

1	2	3	4	5	6	7
2	4	6	10	12	14	16
3	6	11	14	17	22	25
4	10	14	20	24	30	34
5	12	17	24	31	36	43
6	14	22	30	36	44	52
7	16	25	34	43	52	61

You should not read '43' as 'forty-three', short for four-tens and three, because here it means four eights plus three, or thirty-five. With a little practice you could acclimatise to this system. People readily cope with interpreting symbols according to context, reading a digital clock's 16.03 as meaning 'three minutes past four', reading 4/3 as 4 March or 3 April as appropriate, and coping with bases of threes, elevens, twelves, fourteens and sixteens in the antique British system of weights and measures. There is a price to pay for the smaller octal table: multiplication by nine (written 11) needs long multiplication, and so does multiplying by ten (written 12).

Throughout this book I am offering some puzzles which, like Sudoku, depend only on logic and counting. Some are easy; some are not. Here is a first:

GENTLE: Why is $12 \times 12 = 144$ true in octal notation?

Any integer greater than 1 could be used as a base in the same way. With base n, a zero and $(n - 1)$ other distinct symbols are needed. So with base 2, the numerals 0 and 1 suffice. In the binary system, 1001 nights mean not a thousand and one, but nine nights, because they stand for $1 \times 8 + 0 \times 4 + 0 \times 2 + 1 \times 1$.

MODERATE: In binary notation, nine weeks have $1001 \times 111 = 111111$ days.

The binary representation of numbers (actually a seventeenth-century idea) got a good press in the 1950s as brightly modernistic, needed for computers. Constance Reid devoted Chapter 2 of her book to it. Lancelot Hogben's *Man Must*

Measure illustrated binary numbers by using the symbols −
and + instead of 0 and 1, connecting them with electronic
computer storage. I remember this vividly, because oddly
enough, Hogben's bright illustration of techonological application was the very thing that fascinated me by showing the
Hardy-ish independence of pure number from symbol.

The Sudoku numbers are arbitrary symbols: they could be
0 to 8, they could be letters, they could be coloured squares.
Given a puzzle, you can generate many other puzzles by
permuting the symbols − changing the symbols 123456789 to
(say) 783651294. But a permuted puzzle is the *same* puzzle:
the logic is the same, the work of solving it is the same, and
only the symbols have changed. In *Killer Sudoku* the 81
squares are broken up into subsets, in each of which the sum
of the entries is specified. There is an extra rule, which is that
in each of these subsets, no digit ever appears twice.

The symbols therefore actually stand for numbers and are not merely numerals. It is not always easy to untangle symbols from substance, medium from message. The principle of daylight-saving time relies on the *names* of the times of day retaining their cultural force and outweighing the effect of the changed position of the sun. Prices, house numbers, passwords and PINs, car number plates, all have their mix of purely symbolic and actual meaning. The numeral 1 is itself clearly entangled with its meaning: whether in Arabic, Roman, Chinese, Mayan or Babylonian systems, a single stroke gives a picture of One-ness.

People respond eagerly to brands, images and titles. Mathematics is unnatural to word-based culture, because its substance does not depend on names. In quiz shows, a typical question on mathematics is 'what is the name for a five-sided figure?' But this is really a question about English and Greek, not about mathematics. Numbers have no manners, and when introduced at the party rudely say, 'What the hell does it matter what my name is?' The impoliteness of numbers has already taken its toll on the literary consistency of this chapter. Although its subject has been grandly introduced as One, the number is a shifty character, sliding into a more informal 'one' when in use, and then, when involved in arithmetic, compelling the use of a 1 foreign to English prose.

The inhuman factor

Dividing by 1 has already made its unsociable entrance in defining the prime numbers. It is now time for some more steps in what is called the Theory of Numbers, which has its own quite different, and unliterary, elegance.

But first, here is a Richard Dawkins-style mini-lecture on the word 'theory'. It is used in different ways: theory as opposed to something verified, as in a detective story; theory as opposed to practice, as in Marxism or music. But in science a 'theory' may simply mean a *body of knowledge*. In the case of the Theory of Numbers it is just this usage, and in fact the most certain knowledge of anything that we could possibly have. So much for being 'only a theory'.

The prime numbers form a sequence beginning 2, 3, 5, 7, 11... We don't count 1 as a prime, for reasons we come to later. Every number is then either prime, or can be factorised into a product of primes. Thus, $42 = 2 \times 3 \times 7$. This follows at EASY Sudoku level from the definition of a prime; given a number, either it is prime, and we are finished, or it has factors, in which case either they are prime or they break down into smaller factors. However, part of the argument has been hidden in the words 'break down'. It is crucial that if a number has factors then they will be *smaller* than that number: this guarantees that the process cannot go on for ever.

More subtle – in fact a classic of Greek mathematical elegance – is a proof that there is *no greatest prime number,* or equivalently, that there are infinitely many primes. This is decidedly at the TRICKY level, and needs a creative step. First, suppose we have some (finite) collection of prime numbers. If we multiply them all together, and add *one,* we get a new number. For instance, the collection $(2,3,7)$ gives $2 \times 3 \times 7 + 1 = 43$, and the collection $(11, 13)$ gives

11 x 13 + 1 = 144. What we know about this new number is that it cannot be divided by any of the primes in the collection, for if we attempt the division, we shall obtain a remainder of 1. So the new number is either a prime, larger than any in our collection, or it has prime factors, which are different from the primes in the collection. (43 is prime, whilst 144 has prime factors 2 and 3, which were not in the collection.) Now, if there were only finitely many primes, we could take the whole lot of them as our collection. But this would contradict what we have established. Therefore there are infinitely many primes.

More awkward – perhaps even FIENDISH – is the proof of the *unique* factorisation of numbers, an important Oneness of numbers. It is not obvious that the process which tells you there must be *some* factorisation will always give the *same* factorisation. Suppose you are told that

145703 x 8473181 and 91457 x 13498889

are both prime factorisations of the same number. You can check directly that this is not true. (The numbers being multiplied are indeed prime, but the first product is 1234567891243, whilst the second is 1234567891273.) But could you check that there could never be different factorisations, even for numbers with millions of factors, all with millions of digits? It is not so easy.

Here's a smart way to approach the question. If those numbers were really the same, then we could subtract 91457 x 8473181 from both of them, to obtain:

$$(145703 - 91457) \times 8473181 = 91457 \times (13498889 - 8473181)$$

i.e. $54246 \times 8473181 = 91457 \times 5025708.$

If true (which it is not!) this would give a *smaller* example of a number with two prime factorisations. A little thought shows that given any putative example of a number with two prime factorisations, you could *always* derive a smaller example in this way. So there can be no smallest example. Therefore there are no such numbers. This is called the *Fundamental Theorem of Arithmetic*.

You might argue that 1 itself should be called a prime number, since it cannot be broken down. This question illustrates the insignificance of names. If you liked, you *could* use the word 'prime' to include 1, but then the Fundamental Theorem of Arithmetic would say that every number has a unique factorisation into primes-other-than-1. It is purely as a matter of *verbal convenience* that modern mathematicians have chosen to define 'prime' as they do. The numbers themselves are as indifferent to names as to branding, makeovers and sexing up.

Fashion models are not discrete

The numerals, as names of numbers, might be compared with the line and blobs of musical notation. People who naturally play and compose by ear can find it difficult to see the blobs as music, whilst those who have been taught to play from written scores are lost without them. Does a piece of music consist of the score, interpreted by performers, or does it consist of a performance, to which the composer's notes are a guide? A composed score is basically an elaborate sequence of numbers, representing pitches and durations, with the blobs of music notation as its numerals. But a performance is a continuous physical effect in space and time. Music involves a *discrete* abstract plan and a *continuous* interpretation. Music obliges counting, but also *measuring*. Those arbitrarily variable volumes, pitches, and time-lengths need the *continuum*.

The relationship between counting and measuring opens up much more difficult aspects of mathematics, which bedevil those unloved school lessons. It's worth looking at the edicts emanating in 2006 from the Secretary of State for Education, in the light of the current panic over the unpopularity of maths in schools. One dictated that multiplication tables must be learnt by eight-year-olds: for some reason this rote-learning chore is held to be of vital significance. (Using the binary multiplication table would make it much easier to attain: a ritual chant of 'one one is one' would suffice.) But in order to make school maths cool, with 'fashion, football, and the Olympics' as key themes, the Education Secretary stated that a suitable school problem would involve 'youngsters being asked to design a dress and then estimate how many yards of material will be necessary to produce it for, say, 100 girls'.

This surprises me. Multiplication by 100 seems very easy, and hardly requires any more than understanding the way numbers are written in base ten, or the relationship of a penny to a pound. But designing a dress, from a mathematical point of view, seems extremely difficult. Since the whole question is now also one of considerable religious controversy, it would be better to take the simplified fashion problem of boys' trousers: where best to draw the line in deciding how much of the top of the bottom to show. Even this simplified problem is still one of finding the maximum of a continuous function:

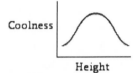

This calls for the much more advanced mathematics of the differential calculus. Anything to do with using figures for measurements brings in questions of approximations, in which the continuum is *modelled* by the discrete. It brings in errors and difficult questions of judgment, going well beyond the counting of One to Nine. These are difficult at school, and remain difficult in the most advanced questions of climate prediction. I am not sure that the education minister had completely thought out what was involved in the mathematics of fashion, football and the Olympics.

Civilisation and language cope with the continuity of the real world by dividing it up arbitrarily into discrete sets. Integers claim to distingush child and adult, for purposes of alcohol, sex, driving, killing. Astrological signs, or the ennea-gram – another fascinating One to Nine – claim to define personality types. Integers rule, from *nul point* in the

Eurovision Song Contest, to pass-or-fail drug tests, to parliamentary votes. Critical phenomena – like a ball hitting the crossbar of a goal – squeeze a spectrum of continuous possibilities into discrete win-or-lose compartments.

Digitalisation has added excitement to this relationship. Music has a new relationship with the integers, because the sound of a performance can be captured on a CD or DVD by what is effectively a gigantic integer. The fact that these integers can be *copied* with amazing ease and reliability gives rise to all the associated problems of communication and copyright. But underneath the simplicities of digital files, there always lies a complex question of the relationship with the continuous or 'analogue' world. Moving images, for instance, depend on an illusion based on the speed at which the brain can process them. The opticians Vision Express insist that eyes register 24 million images in a lifetime. This seems an oddly low figure – about one image for each waking minute – but a deeper problem is how visual images can be discretely counted at all.

In 2006 one of the best-known examples of One to Nine was abandoned. It had long been known that the so-called ninth planet, Pluto, had no better claim to planetary status than Ceres, the largest of the asteroids, and the first to be identified. The fuss about the reclassification, dropping Pluto as a planet, but defining a new class of smaller objects, showed how much names matter to people. In fact the observation of Ceres on Day One of the nineteenth century, 1 January 1801, could be regarded as the beginning of the modern understanding of the Earth's history, with its wide spectrum of collisions from catastrophic to imperceptible. The solar system has big planets, little planets, large asteroids, small asteroids, asteroids breaking up like Shoemaker-Levy, big rocks, tiny rocks, dust, molecules, protons, with no particular place to draw discrete dividing lines.

The relationship of the discrete to the continuous becomes even more problematic when this thought is pursued further. At the atomic level, particles need a *quantum* description. Centuries of puzzling as to whether light is wave-like or particle-like – continuous or discrete – were resolved in the twentieth century by a new kind of description which involves *both*. The year 1900 is important for the definition of the quantum as a unit of existence – a discovery about the number One, whose meaning is still far from fully understood.

Pedants say that 'less' is to be used with measured quantities, 'fewer' with those that are counted. But this distinction is not always clear, even in ordinary life. Do you get fewer bangs for fewer bucks or for less bucks? As extinctions continue, will the oceans have less fish or fewer fish? And looking on the dark side, will we have less light, or fewer photons?

Singles bar

Mary Ann Singleton, who starts page 1 of the *Tales of the City*, reminds us of yet another word root for One-ness; the Latin *semel* that gives us simplicity. This is not the One-ness of counting or measuring objects, but the abstract One-ness behind the ideas of unity, integrity, and consistency –

concepts as important in mathematics as in life.

Being single is not so simple (even though in German, the word *einfach*, one-fold, means 'easy'). I am not just thinking of the common confusions over the plurality of English *data, media, criteria,* or the arbitrariness that makes French *la politique* singular, *les mathématiques* plural. More genuinely ambivalent is the singularity of collective nouns. The public that believes what the government says, is both one and many. *E pluribus unum*, the United States is generally singular but sometimes 'these United States' are more grandly referred to. Yet the Nation of Nations is not united, nor is, or are, the United Nations. John Bolton, the former US ambassador to the UN, and its ex-secretary general, Kofi Annan, have both said, for different motives, that there is no such thing as the United Nations.

'The cultivated banana' is singular, 'cultivated bananas' are plural (although in fact all cultivated bananas are genetically identical, giving the singular a more concrete meaning). Abstraction is counted differently in different languages: *The line of beauty*, in French, would go further to suggest the singularity of *la beauté*. The title of Alan Hollinghurst's prize-winning novel alludes to the line of coke and the (male) bottom line as well as to Hogarth's aesthetics. Which is singular and which are plural, the drug, the sex, the rock or the roll?

Higher abstractions raise greater difficulties. The most elaborated claim for simplicity lies in monotheistic religion, distinguished by the idea of the *one and only one god*. In fact, this subject must have given rise to more human thought on One-ness than anything else. Islam asserts existence and uniqueness with particular insistence, and Islam is perhaps closest to mathematics, with its images from the natural world, and its geometrical patterns. Christianity, with its anthropic images, has involved enormous conflict over the meaning of *one*. The Nicene Creed, committee work intended by Constantine for political unification, skated over the full difficulty. It was left for the Athanasian formula – that the three persons of the Trinity are not three incomprehensibles, but one incomprehensible – to define the distinctive Christian problem with singularity and plurality.

In Bach's Mass in B minor, the Nicene *Credo in unum Deum* is based on a theme which emphasises the number One: Bach reached back into medieval music for a unison chant predating polyphony, and also used the antique notation of the *breve*, virtually extinct by the classical period, as the fundamental unit of time. (Originally the breve was a *short* note, as its name indicates, but by Bach's time a sort of musical inflation had made it unfeasibly long.) Yet even Bach's staggering choral and orchestral counterpoint could not make sense of that word *et* with which soprano and alto open the next section, and so coolly contradict the *unum*.

Fundamentalist mathematics

Even leaving aside the problem of the continuum, there is a deep problem posed by the integers. It is that problem of their simplicity, integrity or consistency – their logical Oneness. How can you be sure that alleged proofs, for instance of the infinite number of primes, are fool-proof? How can you know that a long and difficult proof will never lead to the conclusion that 1=3? In human affairs, apparently authoritative 'proof' can be overturned, as countless miscarriages of justice testify. If mathematical argument is a human invention, why should it not be equally fallible? On the other hand, if it is something above and beyond the merely human, what are the superhuman features that guarantee its consistency, even for statements about infinitely many numbers?

The dream can be traced back at least to seventeenth-century Leibniz, who hoped to find in mathematics an unassailable logical structure for all knowledge. But it surfaced most prominently at the beginning of the twentieth century in Bertrand Russell's work. Russell hoped to derive the numbers from something simpler: logic. This meant finding a logical definition of One to guarantee unity and consistency.

Sudoku gives an analogy. A Sudoku-setter must create a puzzle with a selection of numbers allowing *one and only one* solution. This gives rise to questions which are more difficult and more abstract than that of solving a particular Sudoku puzzle. What is the smallest selection of numbers that will have a unique solution? The answer to this question is currently unknown (it is at most seventeen). How can one be sure that the given starting numbers are sufficient to solve the puzzle? How can one be sure that they do not give rise to an inconsistency? Both these things must be settled to create a valid puzzle. In practice, a setter can simply work through the puzzle step by step and ensure that every square can be filled in uniquely. But mathematics is like an infinite Sudoku, with no way to fill in all the squares. Logicians needed a theory that would establish consistency in some other way.

What is the analogue in mathematics of those start-off numbers in a Sudoku puzzle? They are the *axioms,* and a great deal of sorting-out work in the late nineteenth century had arrived at *sets* as the right place to find suitable underlying axioms. At first sight, or noughth sight, the idea of a set is simple: a book is a set of words, and if you open the book, you see what the words are. But the simplicity is deceptive, involving all those questions of collective nouns and abstractions – and worse.

Chapter 0 of Constance Reid's *From Zero to Infinity* referred to this foundation in sets, saying that One is 'the number of all those sets which contain a single member'. But this looks suspiciously like a circular definition. How do you know what 'containing a single member' means, without knowing the meaning of One?

The ingenuity of Russell's definition lay in characterising single-element sets without using the concept of One. Given a set, we can say it is a single-element set if (1) it is not empty and (2) if you ask any question of an element of the set, you will get the same answer whichever element you take. But only after 344 pages of *Principia Mathematica* could Russell define a single-element set in this way. Those pages involved defining such concepts as 'any', 'same' and 'the', which was a far from easy task. The word 'the', for instance, may seem simple, but it conceals an assertion of existence which has to be made explicit. 'The madness of King George was incurable' is probably a true proposition, but 'Mr Blair was loyal to the King of the United States' is not, and neither

is 'Mr Blair was not loyal to the King of the United States',
for there is no such King.

Russell's principle was that 1 could then be defined as *the
set of all single-member sets*. This is not quite what you might
expect. One is not the *property common* to all single-member
sets, but is defined to *be* a set of sets.

Unfortunately, the concept of 'a set of sets' turns out to
be even more problematic than the United Nations. Here is
one way of seeing why. School maths has unwisely dipped
into set theory in an attempt to look modern, and I have seen
it said that 'The universal set is the set of all sets'. But the
concept of 'the set of all sets' is self-contradictory. Suppose
that 'the set of all sets' is a consistent idea. It is itself a set,
therefore it is an example of a set which is a *member of itself*.
So there are two kinds of sets: those that are members of
themselves, and those that are not. Now consider *the set of all
sets which are not members of themselves*. Which kind of set is it?
If it is a member of itself then it isn't a member of itself. Both
possible answers are self-contradictory. The idea of defining
the number One as a 'set of all sets' cannot be used without
resolving this problem.

Russell had ingenious ways of escaping this difficulty, but
they required very questionable further assumptions. One
striking aspect of the theory was that it seemed to be more
complicated than the numbers that it was trying to explain, so
that it did become circular: you could not understand the logical theory unless you already understood what numbers
meant. Without an understanding of One, how could you
speak of 'a' set? This is the key difficulty that was crystallised

in 1931 by the discovery made by the young Prague logician, Kurt Gödel.

Think for oneself

Gödel showed that there was no hope of regarding mathematics as a puzzle like Sudoku. The analogy fails. The difficulty is essentially this. A question *about* Sudoku such as 'What is the smallest selection of numbers that will have a unique solution?' is not itself a Sudoku puzzle: it is on a distinctly higher and different level. In contrast, a question *about* mathematics is itself a question *within* mathematics. The levels are inextricably mixed up. This follows simply from the business of putting all mathematical arguments into symbols. It is not possible to capture the *truth* of mathematical statements in terms of a Sudoku-like puzzle of proving them from starting axioms.

Gödel's argument had a punchline involving a mathematical statement which referred to its own proof. It was closely related to the problem arising from a set being a member of itself. These and many other developments in twentieth-century logic have emerged like a strange joke at the solemn heart of formality. Good jokes depend on seeing something at a higher level, and then the levels being mixed up. 'Hurt me,' says the masochist; 'No,' says the sadist. In *The Life of Brian*, Jesus's alter ego urges the enlightened axiom of thinking for oneself, only for his disciples to repeat obediently, 'We've all got to work it out for ourselves', illustrating the one thing you *cannot* teach by instruction. Parallel intractable difficulties may be involved in real life logic. 'You must conform to our liberal values' is an example.

If Russell's plan had prevailed, there would be a highly technical definition of One in terms of symbols which only a sort of élite Brains Trust could understand. But in fact, the number One cannot be captured in symbols and there is no better definition than what you can see for yourself. It is an open question as to what this discovery means. Gödel himself seems to have considered that this meant human minds, thinking for themselves, could do something beyond anything that could be achieved by a computer. Other leading people say that this is illusory. For fifty years after Gödel's discovery this remained a rather abstruse argument. But in 1979, the computer scientist and philosopher Douglas Hofstadter published *Gödel, Escher, Bach*, and opened it up with a flood of scientific and cultural ramifications, wonderfully illustrated.

One aspect of Hofstadter's work was that he saw the potential of computers in communication, fifteen years before 'New Media' caught on. He swept these central problems of logic into the enormously expanded symbolic systems made possible by the 1970s, giving an early vision of the scale and quality of what would be possible for the public of the 1990s. Computer science has given logic a new home. The logic that Russell pioneered in *Principia Mathematica* is now the background to database querying. The puzzles that he found are alive and well in that home, and an example came at the opening of this chapter. Originally I wrote the words 'This is Page One', knowing that a statement referring to its own page was just asking for trouble, as indeed it was. Computer typesetting, of which Hofstadter was also a pioneer, gives a miniature picture of the possible tangles of logical typesetting.

This technological connection is not what the logicians of 1900 had in mind: they were hoping to give a complete foundation for knowledge and truth, and might be surprised that the twenty-first century would open with this universal ambition largely abandoned.

In contrast, physics has never given up hope of supplying a complete explanation, and has undiminished ambitions for a single 'theory of everything'. A better word for this ambition might be a *monolithic* theory, for 'monolith', Greek for 'one stone', is translated into German as *Einstein*. It is a tantalising possibility that logic will have to come back into the picture, to make true sense of 'everything'. There may yet be something completely new to be found out about One.

Power of One

There is one completely simple and definite thing to be said about One. It concerns that noughth dimension, *time*. Time has just *one* dimension.

It is this which makes the clear separation between past and future. Two dimensions would give something utterly different; not even science fiction could describe it since even the idea of fiction needs a one-dimensional storytelling. That the dimension of time in the real world is just *one*, is a first fundamental fact that a true theory of everything must explain.

Consciousness seems closely related to the one-dimensionality of time: we seem only to be able to concentrate on one thing at one time. Explicitly in music, drama, novels, but implicitly even in painting and architecture, consciousness needs a linear narrative, ordered in time, with a first page. Writing is the business of turning multi-dimensional facts and ideas into a one-dimensional string of symbols.

English uses almost the same symbol for I, the first person, as for One; and its pronoun 'one' stands for the impersonal I. We look after Number One: the sole soul. But the individual's One-ness is not a static quality, but a process, a dialogue, perhaps an intense struggle. Walt Whitman wrote: *One's self I sing, a simple separate person, Yet utter the word Democratic, the word En-Masse.* Real concentration on real mathematics – as with any concentrated artistic or athletic effort – means shutting out many demands of society. But it depends on a social network. It is an intensely individual process and yet, when mathematical ideas are absorbed, the original individuals become anonymous, forgotten atoms in the fabric.

There are dark sides to the En-Masse, and ideals of Oneness can bring them out. When governments make a disastrous mess they call for national unity: all criticism must cease. The word 'united' too often suggests the dead body over which some past takeover was forced through. The 'United' Kingdom (so flatteringly considered as Airstrip One by George Orwell), arrived with the asteroids in 1801 to crush Irish identity as that of Scotland had already been. The plurality that is Germany has suffered even more from One-ness, the ideal of unity or *Einheit*. The story ran from the Zollver*ein* (customs union), to *ein* Volk, *ein* Reich, *ein* Führer. (Unity Mitford, one of the few people, with Zero Mostel, to have a number as a name, famously adhered to this vision.) Fortunately the Wieder*ver*ein*igung* (reunification) has been a more mixed curse. The Pet Shop Boys end *Fundamental* with 'One... one... one...' on a menacing monotone, responding negatively to the New Labour project of digital identity cards for individuals, and questioning the whole ideal of the *Integral*.

But One-ness can also carry with it the ideal of *integrity,* as opposed to conventional duplicities. Writing in 1956, Constance Reid may well not have been aware that ONE, Inc. was then the name of a first American organisation for homosexual rights. It was an unloved One in the Eisenhower era, but since then, there have been great changes in the per-ception of what 'to thine own self be true' should mean – and contrary to Jane Austen's supposedly universal truth, it is now acknowledged that a single man may not be in want of a wife.

One-ness, in a certain kind of extreme, also characterises autistic people. There are exceptional individuals with great powers of mental arithmetic, whose minds see a vast range of numbers in a direct, intuitive manner. These amazing feats suggest that there is far more yet to be found out about the relationship of number to the human mind, and the true potential of the individual's contribution.

Over to Two

To thine own self and Walt Whitman's careful words 'one's self', 'I and my Soul', suggest that the one of the I is not a simple one. For some it is in Bach's cello suites, for others in the early Bob Dylan, where a perfect solo conveys self-reflection like a dialogue. Inside the I, there is a You – perhaps too many people. To write I must read as if I were you, but if I were you, I would not be writing. The artless *Guardian* writer let slip that 'the universal reader to whom I imagine I am addressing these words is, in my mind, an arts graduate', and as she says, this points to a problem, one which C. P. Snow failed to solve. I cannot solve it either, but I must lead you on as if I could.

Indian mathematicians have a self-deprecating joke: 'India contributed zero to mathematics.' The joke plays on the unnatural language of zero. Contributing 'zero' is not the same as contributing zero: 'zero' is something! This joke is also the basis of a simpler approach to the logic of numbers. Take Zero to be the empty set, the set containing nothing. Then One is the set containing just one thing: namely Zero. Then, Zero and One give *two* things…

Underlining and highlighting simplicity shows its complexity. Numeral 1, adorned with its peaked cap and plinth, stops being a picture of One-ness and morphs into a –

2

To Be or Not to Be

You too, Brutus? I and you, *ich und du*, are number one and number two. You are my equal, but you are not identical to me. Symmetry is about sameness, but the *breaking* of symmetry is about differences.

A perfect symmetry means that two things are effectively identical. Take a simple Sudoku and reverse it, as if seen in a mirror. The new puzzle is solved by mirror reflections of the steps used in the original problem.

Killer Sudoku shares that mirror symmetry. It has another less obvious symmetry, which is of value in solving it. *The Times* gives a 'top tip' for solutions, which is to 'start with the smallest totals'. This is a poor hint, because you could equally well attend to the *highest* totals first. Low and high are related by symmetry. Any Killer Sudoku problem has a dual obtained by swapping 1 with 9, 2 with 8, 3 with 7, 4 with 6, and adjusting the sum totals likewise: the sum S for an n-cell group changes to $10 \times n - S$. Every tip and rule for Killer Sudoku has its dual: the fact that a two-cell 16 can only be $7 + 9$ has the dual truth that a two-cell 4 must be $1 + 3$. A four-cell 11 must be $1 + 2 + 3 + 5$; a four-cell 29 must be $9 + 8 + 7 + 5$. This hidden symmetry in Sudoku gives a picture of the symmetries to be uncovered in the infinite puzzle of the numbers.

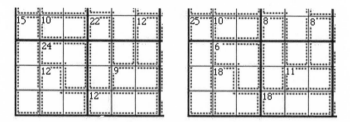

Part of the Killer Sudoku from Chapter 1, with its dual.

A *broken* symmetry arises when there is enough common structure to make a comparison, and that comparison reveals differences. To consider Two-ness is to confront broken symmetries in a world crammed with them. The first is that of yes or no, true or false, the life-or-death dichotomy of Shakespeare's famous two-letter words.

Right and wrong

Considering how galling it is to be wrong about anything,

science is quite good at recognising error and even the eating of hats. By definition, almost anything a scientist says will be superseded by a later and more complete theory. Science is a dynamic process, and it is not only about observing and interpreting data. 'The popular view that scientists proceed from well-established fact to well-established fact, never being influenced by any unproved conjecture, is quite mistaken.' So wrote Alan Turing, the British mathematician of whom there will be more in later chapters.

Yet Turing's own *mathematics* might seem to epitomise the dogmatic yes-or-no 'fundamentalism' that its critics ascribe to science – and which, indeed, scientists may hold against mathematics. It lacks the flexibility of all human statements, with their irony, ambiguity, hypocrisy and deceptiveness. It leaves no room for drama-documentaries, press releases, and other grey areas at the margins of fact and fiction; no room for how you feel about Two, for a new Two, for improving the

 image of Two, or a more inclusive Two. The yin-yang of the Korean flag, with its many symmetries and broken symmetries, suggests a world of subtle complementarities, but mathematical symbols are defined so as to make dualities as aridly exact and inflexible as possible. 'All is not lost', people say, but the mathematical NOT must precisely negate the 'all', not the 'lost'. That NOT (NOT A) is the same as A also requires this peculiar precision. Alan Turing suggested explaining it by saying, 'It's like crossing the road. You cross it, and then you cross it again, and you're back where you started.' But in real

life, we often can't get no such logical satisfaction. Two wrongs make no rights, and doing those things we ought not to have done is not the same as not doing those things we ought to have done.

The OR of mathematics must be identified precisely as the 'inclusive or': A OR B means 'A or B or both'. An 'exclusive or', as in the either/or of 'To Be or Not to Be', needs a different symbol, usually written XOR. Sudoku logic has this precision: a four-cell group that sums to 28 is (9+8+6+5) XOR (9+8+7+4); if it is not one then it must be the other.

The world of propositional logic, which defines this either/or dichotomy, is a small precise world of Two-ness, without the ambitions or contradictions involved in defining infinite sets. But its questions are far from trivial. For instance, how many independent logical relations are needed? We could dispense with XOR by defining A XOR B as (A OR B) AND NOT(A AND B) or (A AND NOT B) OR (B AND NOT A). OR itself is redundant because A OR B can be defined as NOT(NOT A AND NOT B).

DIFFICULT: NOR can be defined so as to generate all of NOT, AND, and OR.

In computer science, the truth or falsity of a statement is stored as a Boolean variable. The term comes from George Boole, an extraordinary individualist of Victorian England who found ways to encapsulate logical arguments that had escaped Aristotle and Leibniz, the defining figures of earlier epochs. Computer 'logic circuits' are based on the essential duality he codified, and computers also exploit the fact that arithmetic in base 2 is very close to the logic of NOT and AND.

It might save a lot of trouble if such logical exactitude were used in legal language. But an either/or mentality is not so obviously helpful to more constructive human thought. The way computers work (or don't work) accentuates the impression that human minds must be doing something quite different. 'To Be or Not to Be' is not about logic but about life and death. Boole called his logical calculus the 'Laws of Thought' but the love (or unlove) of life seems to know no such binary laws. How can there be anything imaginative or creative in a world of pure right and wrong?

Sudoku-solvers will already have part of the answer: there is an art to asking the right questions. The starting point lies in asking 'Where can the nine go in this row?' and 'What can be in the ninth square?', but more sophisticated questions emerge. Advanced guides to Sudoku explain quite elaborate theorems in propositional logic, aided by the efficiency of the two-dimensional eye in pattern-spotting. Chapter 1 has already shown a more serious example of such creative discovery. Gödel's argument was entirely based on either/or logic, but showed something completely new and unsuspected.

Logic itself is open to exploration. The logic of *intuitionism*, for instance, rejects the classical assumption that A OR (NOT A) is always true. It has a 'Not Proven' verdict as in Scottish law. Intuitionism is closer to the idea of actually doing something to prove a result, rather than the Platonic idea that 'it is so'. *Multivalued* logic can embody options like the spectrum from Strongly Agree to Strongly Disagree on tiresome questionnaires. *Fuzzy logic* – as in search engines unfazed by spelling errors – refutes the claim that computers are incapable of coping with grey areas.

The huge area of *probability and statistics* should also give the lie to any claim that mathematics is concerned only with certainties. It can measure degrees of uncertainty. And speaking of uncertainty, twentieth-century mathematics has opened up a completely new picture of physical reality, shaking up logic, number and the concepts of being and not-being. This, the theory of *quantum mechanics*, is intimately bound up with the number Two. But first, we shall look at some more approachable dualities.

Parent powers

Who do you think you are? This is not a reference to the *cogito ergo sum* question of One-ness, nor to *l'Être et le Néant*, but to something altogether more concrete. The BBC series of this title, stimulated by the now enormous genealogical resources available on the Web, has made a great success of presenting the powers of Two.

This is an opportunity to introduce a convention that time always runs up the page, so that some significant events look like like this:

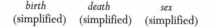

birth   death   sex   party
(simplified)   (simplified)   (simplified)   (simplified)

A typical genealogy, drawn this way, is a tree with ancestral roots:

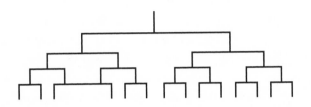

Even ignoring the problem of tracing cousins, genealogical research quickly gets out of hand if every root-line is followed. Two parents, four grandparents, eight great-grand-parents and so on with 2^{n+2} great^n-grandparents, add up to an ever-growing problem. The index n here is a power or 'exponent', and the proliferation of ancestors is an example of exponential growth.

Go back some eighty generations, to the year that ought to be numbered 0, and it is obvious that the world did not have room for your 2^{80} great^{78}-grandparents, for that is about a trillion trillion people. The solution to this puzzle is appar-ent from the genealogies of royal families, where cousin mar-riages of various degrees are commonplace. We are all inbred in some more remote and more common way.

Exponential growth also places in question the antiquated notions of 'a family' with 'a bloodline'. There has recently been much interest in 'holy blood' – direct decendants of Jesus – but after 80 generations, ancestral DNA is diluted to one part in 2^{80}, as meaningless as a homeopathic remedy. In practice, common usage relies on the patriarchal concept of sonship being the bloodline. It is indeed the carrier of the Y chromosome, but this carries only 23 proteins. If Jesus existed, and had DNA, then there would exist cousins of some degree today, about 80 times removed, who would effectively have as much shared DNA as any direct descendants would. More realistically, DNA analysis, and the Y chromosome in particular, has led to new discoveries about the migration of ancient populations and the human relationship to other primates.

For the broken symmetry between XX and XY chromosomes, the reader is referred to literature and to religion, which applies enormous attention to the various anatomical differences of detail. The dichotomy is actually not as clear-cut as it superficially appears, for one thing because of the many gender assignment operations performed at birth. Instead, we will look at the much greater asymmetry in the vertical direction: time.

Call it a loan

The one-dimensionality of time means that the present divides the future from the past. But 'the' past and 'the' future are puzzling concepts, depending on awareness of an ever-changing present. Historians must remember that what is now the familiar past was once the unknown future. Likewise I must remember (with the moving Now approaching the deadline) that what I am now choosing to write will soon be in the unchangeable past. Yet the asymmetry of time does not depend on choosing a Now moment. To an eternal mind seeing all time at once, the direction or *arrow* of time would still be distinct. This arrow is naturally connected with the idea of undoing, inverting, or reversing a process, which is one of the most basic dualities in life and in mathematics, and generally a broken symmetry.

To undo multiplication, we must divide. The idea of fractions allows for any natural number to be divided by any other. The inverse of addition is subtraction. But a large number cannot be subtracted from a small number. The Lord giveth, and the Lord taketh away, but even the Lord cannot take away that which was never given. Instead, the Lord can call it a loan, and set up a debit account. Defining *negative numbers* on this model creates a new symmetry of plus and minus. Indian and Arab mathematicians led the way with a negative symbol.

Again, Europe was reluctant to follow suit. It was not until the eighteenth century that negative numbers entered wholesale into mathematics, and it is hard to understand the reluctance. European culture was full of credit and debt, positive and negative. *Thou art weighed in the balance and found wanting!* – the Good Book is full of Good Book-keeping, with an emphasis on the negative attributes of the ungodly and their fate. Dante's vision of the broken symmetry of sheep and goats is a picture of the integers from –9 to 9, in which Brutus languishes at the negative end. 'Redemption' is Latin for buying back, undoing a debt. It is naturally written $-(-1) = 1$ using negative numbers. But it seems that writing '–1' proved an even more difficult step than writing 0, and confusion persisted until the nineteenth century.

Negative money, though real enough as an overdraft, does not come in the form of coins. It would surely be more blessed to receive negative money than to give, yet one might, if sufficiently unethical, be tempted to throw negative money away when no one was looking. Alas, such greediness seems bound up with the desire to make the future better than the past, and so brings us back to the puzzle of the asymmetric arrow of time.

Law number two

A film shown backwards (as used in special-effect sequences) shows apparently miraculous events. The dead arise, umbrellas are found. Water spurts from the plughole, fills the sink, and neatly aims itself at the tap. What is the reason for this asymmetry?

The reason for the virtual impossibility of such events runs contrary to intuition. It is not because 'gravity pulls water down'. At the level of individual water molecules, every motion could be reversed. The impossibility arises because the large-scale splashiness of the water leads to energy being dissipated as random molecular motions – heat – in your sink. The Second Law of Thermodynamics is the principle governing the asymmetry. It says that disorder increases with time: that on the large scale, the dissipation of heat is never seen in reverse. Strictly speaking, the Second Law speaks of probabilities, but the larger the system, the closer to certainty it is.

Conversely (although not usually put this way) the Second

Law says that order increases in the past, and so must be greatest at the outset of the universe. The night sky looks dark; yet deep into the infra-red it is glowing with radiation left over from its early explosion. The glow from that era is now investigated in detail by space telescopes, and its ordered uniformity is indeed a major puzzle. The totally different nature of Beginning and End to the universe is still a major mystery of Two-ness.

It is hard to reconcile the dualities of youth and age, innocence and experience, with the song of the Earth that physics sings. If disorder increases, how can genetic information evolve from primitive to more complex organisms? How can an individual increase knowledge – or money? For some reason the Second Law is often treated with particular solemnity. C. P. Snow used it as a touchstone of scientific literacy, and Arthur Eddington, who did much in the early days of relativity and quantum mechanics, put it on a pedestal of undeniability. Richard Dawkins recently repeated what Eddington said. Yet it is a very odd law, unlike anything else in physics – literally odd, dividing 2 into an asymmetric 1 + 1. Given that it seems to contradict everything in can-do philosophy, I am surprised there aren't more neo-con sceptics.

Sceptics can be answered: the Second Law applies only to closed systems, but plants and animals are not closed. Life dices with heat-death, but manages to keep going by one chemical contrivance after another, cunningly getting rid of disorder or *entropy* by sending infra-red rays into space. The concept of temperature is closely bound up with entropy, and that is one reason why the question of global warming is not easily settled by common sense, nor just by measuring quantities of energy. As a matter of fact, the problems usually described as energy crises are more properly described as entropy crises. The Earth receives copious energy from the Sun; the difficulty lies in getting it into a form which is useful for human life, which means getting rid of entropy. Unfortunately, it appears that in creating this text, at the cost of computer power used, trees felled and coffee beans burnt, I am throwing out that extra CO_2 which will interfere with the dumping of entropy into space.

Temperature is often used to give examples of how to use negative numbers, since one of the few places where they come into common speech is to express the iciness of the weather. But this makes a poor example of the duality of plus and minus. It depends on conventional scales, where a zero is determined by freezing water (C) or freezing salt water (F). These artificial zeroes do not mark any actual nothingness. The real absolute zero of temperature is at $-273°C$, which corresponds to a (notional) complete absence of motion. Absolute temperature can have no negative value.

By the end of the nineteenth century it was possible to use this concept to tell us what was happening to the temperature of the Earth. Consequently we could deduce that 'greenhouse gases', which prevent infra-red rays going out into space, made it about 30 degrees warmer than what it would be without them. In 1896, Svante Arrhenius, a Swedish chemist, estimated that doubling the CO_2 content of this blanket would increase ground temperature by around 5 degrees. Although this is in basic agreement with modern predictions, it is an object lesson in how science can change its mind in the light of better evidence. For the first half of the twentieth century, Arrhenius was held to have made an erroneous prediction. Experiments suggested that the atmosphere was already as opaque to infra-red as it could get, so that adding more CO_2 would make no difference. Only in the 1950s, using computers, was this assumption shown to be unjustified. It was also expected that the oceans would absorb any surplus CO_2, and more generally assumed that the Earth's system would keep itself in balance. Again, only detailed observation and computation in the 1950s showed these assumptions to be unjustified.

It is a natural assumption that everything has its opposite in the great yin-yang of Nature. But it doesn't – there is no negative temperature. Mass is similar: all matter has positive mass. Nor is there such a concept as anti-gravity. But there is one perfect duality in Nature, to be found in the symmetry of *electric charge*.

Litmus test

The + and – on the ends of batteries are not just convention-al signs. Electric charge gives the simplest example of where positive and negative numbers directly represent physical quantities. If you ask how anything can be less than nothing, an electron seems to say 'look at me!' The duality is exact: like charges repel, opposites attract, in a way that would satisfy fans of the strictest kind of nuclear family without a flicker of bi-curiosity. The duality of electrical + and – also supplies a natural binary storage medium for computers, giving another reason for using base-2 arithmetic.

Electric charge is not usually directly experienced. Shocks of static and bolts of lightning are unusual in showing its raw strength. Everything about chemistry depends on the electric charges in the atom, but it has equal numbers of positive and negative charges and their effects mainly cancel themselves out in the large – exactly because + and – cancel to give 0. We experience only indirect, second-order consequences.

Those indirect effects include the transmission of energy from the Sun, and our ability to see, because light consists of *electromagnetic waves*. Infra-red waves emitted by the Earth's surface, and our bodies, are the same but of longer wavelength. They carry a greater ratio of entropy to energy, and that, quite counter-intuitively, is how life manages to struggle on. Even longer waves exist: these are radio waves, now back in fashion as wi-fi and wireless connections. Shorter waves than those of visible light arise as the sun-burning ultra-violet, and even shorter are penetrating X-rays.

The term 'electromagnetism' arises because the great simplification achieved in nineteenth-century physics showed that electricity and magnetism turn out to be two aspects of same thing. This is a good point to introduce Carl Friedrich Gauss (1777-1855), whose name lives on in the phrase 'degaussing of ships'. Degaussing is getting rid of magnetism, but there would be a good use for the inverse concept of 'gaussing'. For once Gauss had touched something, it became a lodestone for the modern world. I have already mentioned the observation of Ceres in 1801: it was Gauss who computed its orbit and who helped devise the world's first telegraph at Göttingen in 1833. The full unification of electromagnetism and light remained a task for others, including Faraday, Maxwell and Einstein, but it was Gauss who first gave it mathematical life.

Within the unified theory of electromagnetism, there is a subtle symmetry between its electricity and magnetism; this is not at all obvious in the ingenious coils of motors and dynamos, but comes out much more clearly in light waves. It is a broken symmetry, because electrons and protons have electric charges, but nothing we know of has a magnetic charge. And the symmetry is not a simple swap of electric and magnetic forces. You must swap them and then reverse the positive and negative of electricity. This appears as a cycle $(B,E) \rightarrow (-E, B) \rightarrow (-B,-E) \rightarrow (E,-B) \rightarrow (B, E)\ldots$ when the conventional letters B and E are used for magnetism and electricity. This gives an enhanced slant on the duality of BEing or not BEing, but as this depends on the advanced idea of 2 x 2 it must wait until Chapter 4 for a further comment.

When cleaning your teeth in the time-asymmetric sink, you can reflect that your electric toothbrush depends upon yet another broken symmetry. This is that the universe is not symmetric between positive and negative charges: protons carry positive charge and electrons carry negative charge, but they are completely different; in particular, protons are nuclear particles 1836 times as massive as the non-nuclear electron. In wires and motors, chips and screens, the positive-charge protons are stuck in their atomic nuclei whilst the negative-charge electrons are free to wander. But that asymmetry of protons and electrons tells a deeper story.

The electron has an exact dual, the positron or anti-electron, which has positive charge – but a positive mass, the same as that of the electron. Likewise there is an anti-proton with negative charge. These and other anti-particles can make anti-atoms. There is nothing, in principle, to prevent the existence of anti-toothbrushes which would use electricity exactly the other way round (though they would only work on anti-teeth, otherwise there would be gigantic explosions). But the universe we see in the large is completely dominated by matter of one type – electrons and protons, not anti-electrons and anti-protons. The reason for this gross asymmetry would seem to lie right back at the origin of the universe, and is just as mysterious as the direction of time.

Uneven-handedness

Because space is three-dimensional, it does not divide into two parts, so we do not see a left half and a right half. Indeed some people don't have an intuitive sense of left and right. But da Vinci types apart, most people would only have to hold up this book to a mirror to be aware that there is an *orientation* to space.

MODERATE: Explain why a mirror exchanges right and left, but not up and down.

If you're a dentist you may be highly aware of a broken

symmetry in those left and right One to Eights, but otherwise, if you look in the mirror while cleaning your teeth, it may not be so obvious. Generally the human body looks remarkably symmetric on the outside, even if you choose to send subtly asymmetric signals for action or passion with hairstyle or piercings. But inside, the asymmetry, especially of brain function, is striking. Left, right, left, right! Walking or marching gives an on-off duality which is different from the monotony of the heart: it is the natural asymmetry from which music starts.

That broken symmetry, with more people being right-handed, is much magnified in so-dominant culture: *right* is associated with the number One, and anyway means *correct,* this being just one of those numerous crusty Indo-European r-g-t words which enshrine rules and rectitude: reign, regulate, directorship, righteousness, royalties (yes please!), *Dieu et mon droit.* At the right hand of the Father: the Christian creed elevates together the two broken symmetries of spatial orientation and gender. Ironically, the language of human *rights* has attempted to redress the balance for left-overs, the second-fiddle worlds of poverty, race and sex.

Again, although an outward and visible asymmetry in plants like twisting convolvulus is easily seen, it is less obvious that glucose, vital to metabolism, is used by biology only in one ('dextrose') form; its mirror image molecule is useless. A walk into Alice's looking-glass world would leave you starving. Deep down, the DNA helix has also maintained its asymmetry over billions of years. Mirror-images of these molecules would have worked just as well but, for some reason, evolution went with one asymmetrical form rather than with the other.

You may notice a possible analogy: in each country, there is a convention about driving either on the right or on the left. (Not quite: for China is one country with two systems. Hong Kong is still on the left while the rest is on the right.) The origin of the choice typically lies in a small and arbitrary historical decision. It makes only a minor difference what the convention is, but as everyone else is using it, you must conform or die. The great biological asymmetry could likewise have arisen by small chance events, and then have become universal by the dynamics of majority rule. But it is conceivable that those events were influenced by an even deeper asymmetry, the asymmetry of fundamental physics.

This may surprise you, for it is far from obvious that the physical world is asymmetric at a fundamental level. Electromagnetism is mirror-symmetric. So is the geometry underlying gravity, and so is the strong force holding nuclei together, which will be described in Chapter 3. But the weak force, so-called, is another story. The name 'weak' is rather – well, weak – for the force which gives the universe so much of its character. The weak force is the transmutation force, seriously interfering with everything it touches. As such it helps keep the Earth hot inside with radioactivity and hence supplies the world's underfloor heating of about 60Kw per square kilometre (slightly more than the world's power consumption). This, however, is dwarfed by 5000 times more energy coming from the Sun, where again the weak force plays an essential part in turning hydrogen into helium. It is the weakness of the weak force that makes this a very slow process, and so keeps the Sun going for billions of years, giving enough time for evolution.

It is a weak but also a profoundly weird force, with an extraordinary property: it is asymmetric. The weak force knows left from right just as definitely as does a helix of DNA. This was only shown unequivocally by Chien-Shiung Wu in 1956. Her experiment, though it has never become a popular icon of science, marks a major step in the understanding of space and its Two-ness.

The weak force is now far more systematically accounted for, but its underlying asymmetry remains awkwardly expressed and without any fundamental explanation. It has another asymmetry, subtler and harder to detect: the weak force is unlike the other forces in knowing the direction of time. This asymmetry does not appear to have anything to do with the Second Law: it is an exact effect at the level of individual particles.

If you look at particle interactions in a mirror, you will see something that can't happen in the real world. If you film them and play it backwards, again you see something impossible. There is another duality we have already noted: that of changing particles to anti-particles. Under this duality, symmetry is again broken. Yet these three broken symmetries fit together: if you apply *all* of them – anti-particles in a mirror, run backwards – you find an unbroken symmetry, at least for the microscopic world of colliding particles. This unbroken symmetry relies on the fundamental properties of quantum mechanics, and the Two-ness embodied in it.

Lateral thinking

If you have ever heard the word 'duality' used in connection with mysteries of quantum mechanics it might well be from something that is not really a Two-ness at all: so-called *wave-particle duality*. This is a much vaguer use of the word, not a two-fold symmetry, but rather expressing the idea that neither 'particle' nor 'wave' adequately describes what happens on the quantum level. There is, however, a direct and fundamental connection between quantum physics and the number Two, and that is the inescapable role of *complex numbers*. I will put forward the modern point of view, rather than the long historical groping towards them. Complex numbers are simply *pairs* of numbers.

If you think of a positive number as a forward march, and a negative number as taking an about-turn and marching backwards, then a complex number can be thought of as corresponding to leaving the road and turning right or left. Complex numbers extend the duality of positive and negative with a *second* duality. They are defined by the following rules for adding and multiplying pairs. The first rule is unsurprising, but the second is highly unobvious:

$$(a, b) + (c, d) = (a + c, b + d)$$

$$(a, b) \times (c, d) = (a \times c - b \times d, a \times d + b \times c).$$

The natural numbers 1, 2, 3... can be identified with the pairs $(1,0), (2,0), (3,0)$... and for these, the new addition and multiplication coincides with what we had before. The same goes for the fractions and decimals: $1/2$ is $(1/2, 0)$ and π is

$(\pi, 0)$. It goes for negative numbers too: so $(-1, 0)$ is the same as -1. But something new happens when the second element in the pair is used. Following the rules,

$$(0, 1) \times (0, 1) = (-1, 0).$$

The pair $(0,1)$ behaves as a *square root of minus one*.

Roots

The most famous roots are in genealogy, where everyone (in the present state of genetic technology) has unique roots in parents and great^n-grandparents. But you may have no children, or may have many children, so the inverse relationship is different. With squares it is the other way round: every number has its square, but not every number has a square root.

The square of 2, 2×2, is 4, so $\sqrt{4}$, the square root of 4, is 2. Likewise $\sqrt{9} = 3$. If you ask your calculator for $\sqrt{2}$, the square root of 2, you will get an answer such as 1.41421356, which we will look at in Chapter 4. But ask for $\sqrt{-1}$, the square root of minus one, and only a highly sophisticated calculator will respond with something better than a blank look and a sad squawk. It is easy to show that there can be no number whose square is -1: the square of any positive number is positive, the square of any negative number is positive, and 0 squared is 0. This looks like the end of the argument. And yet, by going into this second dimension of numbers, allowing these *complex* numbers, there is a square root of -1 after all. It is $(0, 1)$. More precisely, there are *two* square roots, because (0, –1) does just as well.

GENTLE: Find a square root of (–4).

The definition of complex numbers is, literally, an example of lateral thinking in mathematics. Instead of going up and down the scale of numbers, we move *sideways*. Thus the most logic-bound of disciplines can transcend its apparent limitations.

Addition and multiplication, as defined by the formulas above, behave just like the addition and multiplication of single numbers. In fact, the pairs behave, more perfectly than any human coupling, as if they were one number. This extension of number properties is special to Two. There is no such extension to three numbers. There is a special structure for *quadruplets* of numbers, which will appear in Chapter 4, but it does not make them behave like single numbers.

The extension is also perfectly *complete*, in that no further extensions are needed in order to get a square root of (0, 1), and the square root of that, and so on. This is a consquence of the *Fundamental Theorem of Algebra*, to which Gauss made a major contribution in 1799.

TRICKY: Find a square root of (0,1) (Hint: √2 is involved.)

The first and second numbers of a complex-number pair are called 'real' and 'imaginary'. The 'real' numbers are the numbers we started with, by marching up and down a line. The 'imaginary' numbers then arise as the effect of turning to left and right. It is natural to plot the whole scope of complex numbers as an infinite *plane*, labelled by points (*a,b*).

Amongst these, the numbers $(a,0)$ can be called the 'real' numbers, and behave just like the single numbers that we started with, corresponding to real measurements. But the words 'real' and 'imaginary' are just *names*. There is a way of thinking of complex numbers which emphasises that the 'imaginary' parts also have real significance.

Beyond good and evil

The term 'positive feedback' describes situations where an effect accentuates its cause. If ice melts, the increased area of dark water absorbs more sunlight than did the white reflective ice, and so increases the absorption of energy from the sun. Such positive feedback is also expected to arise as warmer soils release more methane and carbon dioxide.

Such an effect, in isolation, and unchecked – never the real situation – implies an escalating, runaway outcome. It increases faster and faster, doubling and redoubling like the ancestors, in fact with *exponential growth*. Microphone feedback squawks and compound interest are other examples of such spiralling out of control. Equally well, however, positive feedback may lead to *exponential decay*, in which an effect diminishes, and so does the rate at which it fades. A simple example is of radioactivity: the more atoms decay, the fewer are left, and so the radioactivity decreases. Just as the genealogy research problem doubles for every 25 years or so that you try to go back, radioactivity halves in a certain time – the half-life. In short, exponential growth or decay behaves as do the powers of 'real' numbers. A saying from the Gospels summarises both exponential growth and decay: 'That unto every one which hath shall be given; and from him that hath not, even that he hath shall be taken away from him.'

But there is another situation of great importance, associated with anything that vibrates or *oscillates*. To-and-fro cycles naturally occur when an effect *counteracts* its own cause. This is negative feedback. A piano wire is like a spring: the more it is displaced, the greater is the tensile force that pulls it back: this restoring force leads to it vibrating. This is typical of behaviour near a *stable equilibrium*, which a piano wire exemplifies. It is the self-regulation once believed to hold for the state of the Earth. A genuine example in the context of climate change is that increased air temperature may mean more humid and so cloudier skies, which then shade the Earth (although the actual effect of clouds seems much more complicated). If such a negative feedback were the *dominating* feature of the atmosphere, our dear Gaia could look after herself. (The formation of rock carbonates gives a negative feedback on a geological timescale, but that will not help the coming century.) The powers of $(0,1)$ run in a cycle: $(0, 1)$, $(-1, 0)$, $(0, -1)$, $(1, 0)$, $(0, 1)$... and for this reason, any kind of rotational or vibrational motion is naturally described by the powers of 'imaginary' numbers.

This positive feedback leads to exponential growth or decay, and is associated with the 'real' axis. Negative feedback leads to oscillating cycles, and is associated with the 'imaginary' axis. Both effects can occur at once, and that corresponds to the pairing of real and imaginary parts in a complex number.

Economists refer to positive feedbacks as vicious or virtuous circles, depending on the desirability of the outcome. From the precarious standpoint of a polar bear, the melting of the ice is unwelcome, but it is an ill wind that blows no one any good and a sunny issue of *Newsweek* in April 2007 hailed the business opportunities arising from global warming. Compound interest on your debt may seem a vicious thing but to your bank it is of great virtue. Radioactive decay is a virtuous circle if you are hoping to dispose of nuclear waste, but if you were counting on polonium to poison your enemies, I suppose you would consider it vicious. But moral judgments of vice and virtue do not enter into the mathematical concepts. They have no axes of evil, square roots of all evil, or arcs of extremism. The numbers lie beyond good and evil.

Returning to reality, there is a non-obvious but pervasive duality between growth and rotation, and this is a first secret of Two. Take a soap film between two parallel circular rings, and stretch it out. Its shape illustrates the duality perfectly. Between the circular rings its shape is that of an exponentially increasing curve.

In the nineteenth century many such physical effects were encapsulated using complex-number methods. They were ready and waiting, thoroughly Gaussed, for the twentieth century. After 1900, quantum mechanics revealed that complex numbers do not just supply a method for pairing numbers and properties together; they are the actual medium in which the world works.

Reality, but not as we know it

Quantum mechanics goes right against the either/or logic of common sense. One cannot think of an electron as an object that either is or is not at some point. In fact the ontology — the 'To Be or Not to Be' — of electrons is a highly contentious question. They are described by 'wave functions' or 'quantum states' which are nothing like everyday objects. It is not even valid to speak of 'two electrons' with the usual sense of Two-ness applied to counting two separate things. There is no way of telling which is which; they are 'entangled' and form something more like a Two-ness of electronicity. Strictly speaking, all the electrons in the universe are entangled in a field of electronicity. But the mathematics of complex numbers can successfully cope with describing this breakdown of ordinary logic.

The strangeness of quantum mechanics is making an impact on practical yes-or-no questions of logic, information and communication. As computers get smaller and faster, they are heading towards a point where their operations will depend on very small numbers of electrons. This threatens the accuracy of the either/or binary logic, because electrons are not either/or entities.

But since the 1980s, much progress has been made towards building *quantum computers*, to take advantage of the very features of quantum mechanics which seem problematic. A quantum computer can use a new kind of quantum-mechanical Two-ness (a 'qubit') instead of the discrete yes-or-no 'bit' of computer logic, in a way that makes it possible to pursue many calculations at once. The related engineering of quantum communication can make constructive use of entanglement to pass information across space in a way that would be impossible otherwise. Quantum Two-ness is real, although it is not reality as we know it.

Quantum mechanics supplies the invisible infrastructure of everything we know. After 1800, vague alchemical guesswork about natures and affinities was turned into crisp laws about how chemicals combine, expressed in terms of integers. From the mid-1920s it was clear where these integers came from: the quantum mechanics of the atom. The essential mechanisms of life likewise depend on quantum mechanics. This was appreciated before the structure of DNA was identified. In 1943, the physicist Erwin Schrödinger pointed out in an essay called 'What is Life' that the One-ness of the quantum makes possible the encoding of enduring, reliable information. This was verified ten years later when the famous double helix was elucidated. The Two-ness of the helix is a quantum mechanical Two.

The nineteenth-century calculation of the effect of CO_2 was done without knowing why it was opaque to infra-red radiation. Later, quantum mechanics explained it as a general property of larger molecules of which CO_2, water and methane are examples. So it is for reasons involving the nature of electrons that global temperature is expected to increase. The fact that quantum mechanics is involved is another reason why the predictions will not necessarily coincide with intuitive convictions about the weather always being unpredictable, CO_2 being good for plants, the Earth being too big for mere humans to affect it, that what goes up must come down, and so forth.

If complex numbers are so intrinsic to reality, why are we unaware of them? Why don't we visualise complex qubits rather than Boolean bits of 0 and 1? The standard answer is that *measurements* break the complex numbers down into classical yes-or-no logic and real numbers. But this is far from being fully understood. How and when do measurements happen, and how are they related to the mind's perception of the world? There seems to be perfect randomness involved in the process, but what is its origin? How is this connected with time direction, entropy and the Second Law? If quantum mechanics is about small systems, how large is small? Schrödinger devised a famous thought-experiment to ask just this: in modern terms it amounts to asking what happens when a qubit in a quantum computer is measured and turned into an either/or bit of information in an ordinary computer. Schrödinger dramatised the situation by imagining using the computer bit to kill a cat – nothing to do with real animal experiments! The question of whether the cat is *To Be or Not to Be* goes to the heart of this open question about Two-ness.

Spheres of influence

To look a little deeper at the Two-ness of quantum mechanics, it is handy to examine something that sighted people use all the time, but do not necessarily ever think about: the Two-ness of the sphere. Visual information comes through the Two-ness of the retina, collecting from a two-dimensional sky of light; nerves fan out to a two-dimensional skin interface with the tactile world, and the brain is largely a folded two-dimensional surface. No more than a portion of the sphere of light can be seen at once, and it is the brain's putting together those portions of the sphere that creates our visual sense of space. It's not obvious why animals have not developed all-round vision, analogous to hearing; if we could see from the backs of our necks we might have a different sense of space.

The complete geometry of the sphere is not so easy to grasp, for 'up' and 'round' are not precise, and East meets West everywhere. Are you sure you know why summer days are longer as you go further north? Can you guess the direction of Mecca from Hawaii? There is a particular problem with time zones at the poles, with choosing an international date line, and defining a truly round-the-world adventure. These difficulties correspond to genuine mathematical problems with describing the Two-ness of the sphere.

As it happens, the Two-ness of the sphere gives the best picture of quantum-mechanical Two-ness. Although I have described complex numbers as if on a plane, with a real and imaginary axis, a simple quantum-mechanical system is actually better described by the geometry of a sphere. This is because it is properly defined by the *ratio* of two complex numbers, rather than by one complex number. This is not quite the same thing and it curls up a plane into a sphere. A qubit of quantum computing can be thought of as a point on a sphere.

Points of view

There is something even less obvious about spheres, which reveals a more secret aspect of Two. We return to the picture of a sphere as the sphere of eyesight. If you adopt a new point of view just by moving your head, what changes can you make in your field of vision?

You may never have noticed that this involves a *three-dimensional space of possibilities*. First, you can move your head to right or left, then move it up and down, to make any direction you like the centre of your field of view. Then (admittedly, with some limitations, unless you are extremely athletic) you can rotate your head while keeping constant this centre of the visual field. This makes three parameters. Some people will be acquainted with this Three-ness as roll, pitch, and yaw, the terms for the three different ways in which a ship or aircraft can rotate, but the more nautically challenged probably lack a name for them.

What is the shape of this three-dimensional space? You are familiar with it from every waking moment; it is the space of your seeing, looking, searching, probing consciousness; you actively navigate it, and yet you are unlikely to have been aware of it needing any conscious description. You are likely to be completely unaware of its shape. It is a finite space, but without any edge, so it is nothing like any three-dimensional object. It actually has almost the shape of a *3-sphere*, which means the analogue of a sphere in four dimensions. In Chapter 4 I shall define this properly, but here we need only one thing, the significance of that word 'almost'. The space of visual possibilities is actually only *half* such a 3-sphere.

Spin

The factor of Two is the inner secret of space, which has only been known since the 1920s, but is vital to the nature of matter. It is the two that appears in the 'half' when an electron is said to have a spin-half, meaning a spin of 1/2 of Planck's quantum of existence. Here is a shoe-string experiment which gives you a way of seeing that factor of two directly.

MODERATE: Tie a pair of shoes to each other with (suitably long) shoe-laces. Now rotate one shoe completely round, so that the laces get twisted too. Rotate again, in the same direction, so the laces are doubly twisted. You will find that you can untwist the laces by passing the shoe between them, without rotating it at all. Only a double rotation can be undone like this.

Electrons are embodied in space in just such a way, as if with strings attached. This follows from their nature as complex-number entities. When given a single rotation, an electron is not back where it started: its virtual laces are twisted. Only when rotated a second time, do they become untwisted. This factor of two is consistent with saying that the electron has spin 1/2. It also gives the electron a double ration of possible states, and this factor of two is the fundamental fact of chemistry. The first place it shows up is in the inertness of helium.

The helium nucleus, emerging at 100 million degrees from fusing two protons together, is vital to the process by which the Sun pours out energy. There is another, gentler, cooler story about Two that comes from investigating helium as a chemical element, complete with its two electrons matching and cancelling the electric charge of its two protons. In the early nineteenth century it was completely unknown. It went undetected on the Earth because it is virtually inert chemically: it bonds with nothing. After the birth of quantum mechanics this complete lack of chemistry could be explained. Helium has a complete set of both of the two possible electron states, one for each state of spin, leaving it with nothing to share. This is the start of a pattern which runs through the table of the elements, and dominates their chemistry.

A new two

To say that helium is inert because it is 'complete' with two electrons makes an implicit appeal to the idea of there being just two places available for just two electrons. This needs to be made explicit.

Electrons cannot be thought of as objects with 'places'. They are in 'quantum states', which do not correspond to anything in ordinary experience. However, there is one aspect of a quantum state which does correspond to what we expect of something called a 'particle': roughly speaking, a state either has one electron in it or none; it cannot have two. There is a genuine *To Be or Not to Be* for electrons, called the 'exclusion principle'. It turns out to be a logical consequence of its spin-1/2, and so of complex numbers. For the helium atom, there are just two possible states of lowest energy, one for each spin, and if these states are filled then there is indeed no room for more.

Protons and neutrons in the nucleus have the same exclusiveness, and it is this that gives matter its hard and unyielding quality. But this is not the end of the story. At the atomic level, *forces* also must be considered as quantum states. Electricity, magnetism and light, when interacting with electrons and protons, must be described in terms of *photon* states. But photons have no such exclusion principle. There can be limitless numbers in the same state: an intense laser beam gives a picture of what this is like.

Particle-like entities are called *fermions* (for the Italian, Enrico Fermi) and the force-like entities *bosons* (for the Indian, Satyendra Nath Bose). The properties of fermions and bosons are very different. Yet their equations differ only through choices of plus and minus signs. One reason for this simplicity is that they are united by Planck's constant, the One-ness of the quantum, which has the same value for both forces and particles. In the 1970s, physicists began to express this duality by a new idea: *supersymmetry*.

Supersymmetry swaps those plus and minus signs. It must be a highly broken symmetry, because we don't see any fermions related to the photon by supersymmetry. The Large Hadron Collider at CERN, the European particle laboratory near Geneva, will soon start to look for evidence of new heavy particles which, if found, would make more sense of it. But at present it remains mysterious, and shows how the logical either/or of mathematics is not the whole story. It is a new Two-ness, and different people feel different things about it. Physicists have used it to give simple proofs which, to mathematicians, don't seem to make complete sense.

The story of Two is not finished. Negative and complex numbers seemed crazy at first. Mathematicians eventually found a consistent logical framework for them, while physicists found them in electricity and quantum mechanics. With supersymmetry, physics seems to be a step ahead. The number Two is not completely understood in itself, and needs a bigger picture of the world. The sphere of sight naturally leads us on to look at –

3

Trinidad

Sanctus, sanctus, sanctus: what I tell you three times is really, very, truly, awesome. Three doesn't just talk; Three asserts. Three is superlative; Three thinks big. Great, greater, greatest: a rhythm of three gives ten, hundred, thousand. A thousand is ten times ten times ten, and Fibonacci taught Europeans to write their larger numbers in thousands. A million is a thousand thousand, but a thousand thousand thousand is – well, just to be confusing, usages of 'billion' have differed from the beginning, with European billions being a thousand times greater than American billions. As in Iraq, the British government has parted company with Europe to side with America in counting 10^9 as a billion, and 10^{12} as a trillion.

Such large numbers have no immediate meaning. Billions, as for instance somehow showered on computer systems, are often confused with mere millions. Douglas Hofstadter has written a superb essay on the difficulty of visualising a billion of anything, quoting an American senator: 'A billion here, a billion there, soon you're talking real money.' (That was long before the US casually dispersed nearly $12 billion in $100 bills as a detail of the 'reconstruction' of Iraq.) Perhaps Three is at the limit of what the mind can cope with directly. One, two, three, many… Gold, silver, bronze and also ran… The logic of number triplets, when solving 3 x 3 Sudoku, is near to the limit of direct perception.

Yet billions are real. Billions measure the age of the planet, the diversity of its life, its human population, and the spectrum of human individuals' wealth and power. A billion heartbeats roughly measure the span of animal life, and a billion words a lifetime of consciousness. I think of this when the word 'gig' is casually used for the gigantic gigabyte storage now routinely available. The K, the meg and the gig give alternative words for a thousand, million and billion, and they tell a story relating Two and Three with Ten. The K for a thousand, the third power of ten, is convenient because it is close to 1024, the tenth power of two.

EASY: This is also why 2^{80} ancestral lines going back to the year dot gives about a trillion trillion.

Less obvious is that the approximation $1024 \approx 1000$ is related to the problem of *music*, and this gives a starting point for both the harmonies and the cussedness of numbers.

The sound of music

The musical relationship between Two and Three is most obvious in rhythm. In the European tradition, three-rhythms for dance music predominated until the twentieth century. Then American marching bands became sexy with ragged-time four-four, and complex African-American rhythms soon changed popular music completely, leaving triple time as a wistful folk-music nostalgia. But there is another less obvious Three-ness in the distinctive European tradition of *harmony*. This is based on playing a chord of three notes at once, and is still very much alive. Which three notes? The answer depends on powers of two, three and five – with three as the dominant number.

Sound is carried through air by waves with frequencies in a range from about 20 to 20,000 per second. The shorter the wavelength, the higher the frequency and the higher the sound. Each musical sound comes from a length of air or solid medium vibrating with some fundamental frequency, or pitch, but that vibration is bound to contain higher frequencies at (some of) 2, 3, 4, 5... times that basic pitch. This happens naturally through the way vibrations arise. The timbre of an instrument is largely determined by the mixture of harmonics it produces. The clarinet, for instance, has no even harmonics.

The harmonics determined by the multiples 2, 4, 8, 16... are very special. They are so closely related to the fundamental vibration that they are interpreted by the brain as being the 'same' sound at a higher pitch. This gives to music a basic, natural connection with the powers of two, and so with the exponential growth as met with the proliferation of ancestors. In fact, some instruments display a sort of graph of exponential growth. It is striking in the shape of a harp, and even more explicit in a row of organ pipes, which gives a direct picture of the actual wavelengths.

The next most simple relationship is given by the *third* harmonic, and this gives the most closely related but different note, called the dominant. Beethoven's Ninth Symphony begins and ends with a primeval sound, as heard in medieval European music, a chord of only two notes. These two notes are determined by the 3:2 ratio. The ratio is, confusingly, called a *fifth*, from counting the five notes CDEFG from C to G. Likewise the 5:4 ratio, based on the fifth harmonic, gives a (major) *third*, the term coming from the three notes counted CDE from C to E. The term 'octave' comes from counting the sequence CDEFGABC.

If further notes are to be combined harmoniously, what should their frequencies be? The principle of harmony is that their ratios should be based on simple numbers, so as to make as few clashes as possible between harmonics. But this is not as easy as it sounds. Imagine a string orchestra tuning up (actually changing the speed of sound in their strings by adjusting their tensions). Suppose the cellos have agreed upon and are sounding their low C, and the violins use it to tune their four strings to G, D, A, E in turn. To tune the G-string in perfect harmony with the cellos, the violinists locate the note which has exactly three times the frequency of the low C. Next, the D-string must have a frequency 3/2 times greater, the A-string then $3/2$ of that and finally the E-string $3/2$ greater again. The five notes C G D A E suffice for all the pentatonic melodies common to world music. But there is already a problem with the harmony, even for this basic scale. For the ratio of the final E note to the original C note is $81/8$. This is almost, but not quite, 10. As a result, it will clash with the *fifth* harmonic in the cello's vibration. It is slightly higher, or sharper. This clash contradicts the principle of resonance on which we based the allocation of notes.

Another problem faces a piano tuner who starts with all the C notes, then tunes each G to be in agreement with the third harmonic of the C. Similarly the tuning can go on to D, A, E, B, F#... 'which brings us back to Doh', as the song goes — but doesn't. The sequence would only return to its starting point if twelve fifths were equal to seven octaves, equivalent to the non-equation

$$3^{12} = 2^{19}, \ 531441 = 524288.$$

MODERATE: This, together with the untruth $80 = 81$, implies the untruth that $1024 = 1000$. This untruth also shows why three major thirds do not quite equal an octave.

MODERATE: The non-equation $63 = 64$ is equivalent to taking the *seventh* harmonic to be given by two perfect fourths.

Fortunately, the Western ear allows itself to be fooled by a compromise with truth. The standard fix needs mathematics beyond fractions, and we shall do this in Chapter 4. Meanwhile, our tour goes from tonality to tones of colour, where Three plays a completely different role.

The eye of the beholder

Light is an electromagnetic wave, and what we call its colour is a measure of its frequency. Light waves travel nearly a million times faster than sound, and the vibrations we perceive are also of much higher frequency than the vibrations of sound, around 500 trillion a second. The spectrum of the rainbow is the equivalent of about an octave. But light has no effect on the eye analogous to the harmonies of sound: two frequencies related by simple proportions do not have any effect on the eye at all. There is nothing in the way that light is created or absorbed that gives rise to an association between such frequencies. The reason for that, as for so much in life, lies in quantum mechanics.

The integer-based quantum does give rise to integer-based patterns, but they are of a quite different nature. It had been noticed in the 1880s that hydrogen emits light with frequencies proportional to $(1/m^2 - 1/n^2)$, for integers m and n. The explanation of this pattern in terms of quantum leaps was one of the first great successes of quantum mechanics. The leap from $m = 2$ to $n = 3$ corresponds to a distinctive red colour. Other vivid frequencies, the analogues of pure musical notes, are seen in neon signs and sodium lamps. But they create no harmonics, and no chords.

Few people possess 'perfect pitch' – the ability to recognise and name the pitch of a sound. Yet sighted people expect of each other the equivalent of perfect pitch for colour, giving detailed names to different frequency effects. These roughly agree, although words and their associations differ with culture and language. English is very clear that pink is pink, not light red; whilst light blue has no name of its own as it does in Russian. Paint colour charts go wild with imaginative names for more subtle distinctions. These names, and the boundarylines between different colours, are bound to be somewhat arbitrary, because colour is *continuous*. The interesting thing is that it is a continuum in a *three-dimensional* space, even though the spectrum of the rainbow is determined by the one dimension of frequency.

The reason is that the retina has *three* kinds of cells reacting to light, based on three different molecules. These peak in three different parts of the spectrum: actually in blue, green and yellow-green areas. Perceived colour measures the intensities with which these different cells react. 'Colour blindness' arises for people in whom these three molecules are not functioning, but it is an odd fact that all human eyes are rather blind to red, even though red is so clearly distinguished in language. It seems surprising that we distinguish red, orange and yellow so clearly, but do not differentiate those blue-green colours to which the receptors are more sensitive. This emphasises that perceived colour is an artefact of eye and brain processing.

The Three-ness of colour is entirely in the eye of the beholder, and this means that the brain accepts completely different combinations of light as being of the 'same' colour. Yellow may come from a sodium-yellow lamp which emits virtually a single frequency. It may come from a mixture of red and green dots on a television screen. It may come from pigments, as in van Gogh's sunflowers, which absorb blues from white light falling on it, and reflect the rest. In fact there are infinitely many ways in which the 'same' colour may result. This has no analogue in sound.

The enormous world of photography and paint rests on exploiting and fooling the three-dimensional geometry of the eye's colour space. It used to be said that the camera cannot lie, but everyone knows that a Photoshopper can tell a whopper. You may frown on manipulating photographs, but the very nature of photography, like painting, is that of deceiving the eye by creating a mixture of light frequencies which make the same impression on the human eye as some quite different mixture. (This may be one reason why animals don't affect much interest in art, television or computers.) The question of which colours can be so imitated, and which cannot, is very sophisticated. The colour spots used on digital screens – which are not at all the same as the colours which mark the peak response of the eyes' receptor cells – cannot in fact reproduce all colours. Somehow the brain forgives a great deal.

Websight

If you write HTML to create a webpage, and use the instructions for getting coloured backgrounds and spaces, you experience a direct introduction to colour three-dimensionality. A colour is coded in HTML by three numbers in the range from 0 to 255, specifying the red, green and blue intensities (roughly as they are perceived by the eye). For instance, black is (0, 0, 0), white is (255, 255, 255), grey is

(128, 128, 128), yellow is (255, 255, 0) and Hot Pink is (255, 105, 180). The colour-space is thus a cube. The discrete values mean that it is not a true continuous cube, but this does not affect the essential point about three-dimensionality.

The reason for the number 255 is that HTML uses these numbers as written in base 16 ('hexadecimal' or 'hex') notation. Base 16 needs some extra numerals to represent the numbers from ten to fifteen, and the convention is to use ABCDEF for these, so the numerals are 0123456789ABCDEF. (The same convention is used in 4 x 4 Sudoku puzzles.) The numbers from 0 to 255 are then simply the hex numbers from 00 to FF.

MODERATE: Hot Pink, by any name, would be the same. Show it is represented by FF69B4. How many different colours are there in the colour cube?

Software for editing images may represent this cube of colour in other ways. Instead of describing a colour by values for redness, greenness and blueness, it may be described by brightness, saturation and hue. *Brightness* measures the sum of the three colour intensities. If you restrict to values which add up to 255, you factor out brightness and reduce the possibilities to a colour *triangle*. This restriction is equivalent to slicing through the cube at the three vertices nearest to 0. The resulting triangle has pure red, green and blue at the vertices. *Saturation* then measures distance from the grey centre of this triangle, and *hue* the angle round the triangle. Experience shows that navigating this three-dimensional space of possibilities is highly non-trivial.

DIFFICULT: Imagine holding the colour cube with white at the top and black directly beneath it. If you divide it exactly in two by a horizontal cut, what shape does the cut surface make, and what colours appear on it?

Fooling the eye with colour *printing* is a more difficult problem, and the standard 'CMYK' solution brings in more advanced geometry. It is the dual, or negative, of the problem of colour on luminescent screens. To produce white on a white page means printing nothing at all, leaving the white light that falls on it to be reflected. This is the analogue of producing black on a screen by doing nothing. Every other colour effect involves a similar reversal. To produce blue on a printed page means that the inks applied must absorb the red and green from the white, and reflect the blue. The red-absorbing ink is *cyan* and the green-absorbing ink is *magenta*. If both of these are printed in tiny dots, they do indeed fool the eye into thinking that it sees blue. The third of these negative colours is supplied by yellow ink, which absorbs blue light. Cyan, magenta and yellow are the C, M and Y dimensions. In principle, ink intensities drawn from just these three dimensions should be able to imitate all colours. However, the physics and chemistry of inks on paper means that a satisfactory black is not obtained by the printing of all three inks at full intensity. So a further black ink is used as well. This makes a *fourth dimension* available. Often, however, the space of colours printed is actually restricted to a *three-dimensional hypersurface* within that four-dimensional space. The location of that hypersurface may be decided by software converting an RGB image into CMYK space, dictating exactly when the black K will be used instead of a sum of C + M + Y.

If we had ultra-violet sensors like bees, colour space itself would be four-dimensional; unfortunately, we cannot read the come-on signals of the flowers, which are designed for bees to see. Many birds have four-colour sight, and we have a very limited idea of what the world looks like to them, or what they see in us.

A model of life

Of course, physical space, the space that estate agents describe as 'location, location, location', is also three-dimensional. A photograph or representational painting is, however, two-dimensional. Both of these imitate what the eye sees, which is also (a patch of) the two-dimensional sphere of sight. The standard method of perspective is a transformation of three-dimensional space into a plane image, keeping straight lines straight. This is also roughly what a camera does. But it is an artifice. It is only 'realistic' from one vantage point.

It is difficult to imagine the power of the *trompe-l'oeil* effect that the first fully worked perspective drawings must have had on their viewers. (Masaccio's *Trinità* in Santa Maria Novella, Florence, around 1427, is a famous example.) But perspective only works satisfactorily for comparatively small fields of view, and never allows what the eye does in reality, which is to look right round. On panoramic photographs, straight lines on buildings become curved lines. These contradict the principle of perspective drawing but are more 'realistic' for the sphere of eyesight. Stand right in front of a long wall and ask if it looks 'really' straight.

Pictures and photographs are *mappings* from a three-dimensional space into a two-dimensional *image space*. Modern mathematics has run roughly parallel to modern art in probing the question of what representation *means*, and extending the idea to mappings involving spaces of any number of dimensions. The conversion between RGB and CMYK is a model of the way that the geometry of higher dimensions emerges as a natural development.

Mathematical spaces can be thought of as general *spaces of possibilities*, with the space of possible colours as a model. The common expression 'another dimension' is quite correctly used to mean bringing some qualitatively different consideration into discussion. Dimensions may also be referred to as *parameters* or *degrees of freedom*.

As an example, the coolness graph for boys' trousers in Chapter 1 was an over-simplification of the fashion problem, by reducing it to just one dimension. In reality, the great challenge for a young man is to decide not only how much bottom to show at the top, but how much of the thigh to hide. Two dimensions are needed for these parameters alone. Fashion is always changing in time, and showing time as another dimension makes this problem of dress design, as urged by the Department of Education, into one of studying a three-dimensional hypersurface in a four-dimensional space.

GENTLE: Assuming that your life-class model sits still, what is the dimension of the space of possible camera shots that you can obtain? If you are also free to move a spotlight, how many extra dimensions does this add to the space of possibilities from which you select your composition?

Although we think of living in a world of three dimensions, the actual space of possibilities in which we have to operate is generally a space of many more dimensions. Consider a single shot in a ball game. At one instant, the position of a ball is specified by three dimensions. But its orientation gives another three – the same three dimensions as we counted in Chapter 2 as the space of rotations. The motion of the ball is then given a further three for velocity and three for its spin. (It is the spin which, as people say, gives a whole new dimension to the professional game.) The problem of batting against such a ball requires predicting how the ball will continue to move in these twelve dimensions, and then choosing a point in a similar twelve-dimensional bat space to optimise the outcome. In a recent cricket match between England and Pakistan, the allegation of 'tampering' with the ball to break its symmetry in one of these dimensions, thus complicating the prediction of its motion, turned into a major dispute.

How can the even greater problems of economics and life be reduced to manageable dimensions? One fascinating development lies in simplifying them to spaces of *strategies* with a small number of parameters. This gives the model of life known as *Game Theory*. The simplest kind of game is a *zero-sum two-person game*, in which a loss to one player is a gain to the other. As it happens, the theory of such games also leads to an example of mathematics in which the Hardy/Hogben dichotomy evaporates. In 1940 Hardy quoted Hogben on 'planning' with distaste, but the practical application of mathematics to strategic planning in the 1940s turned out to require a beautiful example of abstract many-dimensional geometry. Applied to zero-sum two-person games, it involves an elegant theorem using a duality between such spaces, to show how the strategies of the two players are locked into an embrace.

A famous example of such a game, rather easier than cricket to understand, is that of Scissors, Paper, and a third which is Stone or Rock depending on your side of the Atlantic. Each player has a three-dimensional strategy space, and the solution for optimal play is for each to choose the point in that space, which means playing the three options equally and at random. In this case it is fairly obvious that a player can win, and can only win, against an opponent whose choices fail to be random and equiprobable. Mathematical game theory shows how to solve much more general and less obvious strategic questions – though the solution opens up yet another question, of how to ensure randomness.

Trials and tribulations

It is common to applaud 'win-win situations' which arise from games which are not zero-sum. Unfortunately, turning now from optimism to pessimism, they can also be lose-lose situations. If you are a billionaire, you can probably make the rules to suit yourself, but those on the receiving end may be faced with less convenient constraints. The classic example is that of 'prisoner's dilemma', a situation which could well be imagined in a Caribbean setting. Suppose two captured prisoners are interrogated separately. Both have a choice: to inform against the other, or not. They are both best off if both refuse to inform. But the direst outcome for a prisoner is when, having refused to inform, he finds himself incriminated by the other. The only way to avoid this worst scenario is to inform; the same logic applies to both and so both will incriminate each other. Neither can choose the strategy that would be to their collective advantage.

Refusing to inform is called 'co-operating' and informing is called 'defecting'. Of course, from the point of view of the interrogator, the 'co-operative' prisoner is the one who can be made to squeal. This shows why a non-zero-sum two-player game is equivalent to a zero-sum game for three players, in which the players can make and break alliances. (It is zero-sum because one may assume, reasonably enough, that the more painful the outcome for the prisoners, the happier the interrogator.)

The best-known work of mathematician John Nash, of *Beautiful Mind* fame, established that for any such game, with three or more players, there would be an 'equilibrium' situation in which no player could gain by changing strategy. This, however, does not resolve the problem of how to bring about collective betterment in the face of individual risk. One approach is to consider not the strategy for a one-off game, but the strategy for repeated trials in which there is opportunity to learn from an opponent's previous form and react accord-ingly. Extensive experiments have shown that a 'Tit for Tat' strategy establishes a much better long-run outcome. This is based on co-operating, but defecting if the opponent does. Such an extended scenario does make the biological evolution of altruism more comprehensible.

Douglas Hofstadter wrote extensively on this in *Scientific American* and then put his readers to a practical test. He offered a prize in a Luring Lottery; it would be a million dollars, divided by the number of entries. Individuals would be allowed to submit multiple entries. This created a many-person game, the players being the many readers of *Scientific American*. The best outcome (or from Hofstadter's point of view, as prize donor, the worst) would be if there was just one entry. No entries, or very many entries, would be equally bad (or good). How could the millions of readers co-operate so as to submit a single entry? Hofstadter had a super-rational plan for how this could be achieved, but it only needed one defecting reader to wreck it by putting in many entries. He need not have worried that it would be followed. For he was flooded with entries, as readers vied to write down the largest number they possibly could.

In practice, people do co-operate in many ways, despite the lack of individual benefit. (There is, for example, no rational point in any individual troubling to vote in a large-scale election, as the chances of the election being decided by one vote are extremely slim.) But co-operation is fragile, as gun control and gang revenge questions indicate. The nuclear non-proliferation treaty, like other self-denying ordinances (the ABM Treaty and laws of war generally) are similarly unstable. The problem of mitigating climate change is also of this nature, and even more complex because it involves a vast number of players.

Co-operating in this game means making some sacrifice for the sake of those who are not just out of sight, but as yet unborn. But worse, there are plenty of players in the global game who will defect with pleasure, because a warmer climate suits them, because they believe it will advance the End Times, or because they don't give a damn. Many more can at least live happily with predictions made for their lifetime, even if they don't believe in them at all. Science needs a web of trust, quite unlike mathematics, where in principle you can see everything and think for yourself. If a professor of meteorology in a distinguished institution pours public scorn on the predictions of climate change, then inevitably that trust will be diminished.

On the governmental level, the year 2007 seems to have marked a decisive point. The British Conservative party has professed that 'the politics must fit the science and not the other way round'. Even the United States administration is no longer asserting disbelief in the scientific consensus. But science is not easy to fit in with. It will be hard to overcome that kind of 'realism', which means boasting of outdoing others in ever-growing wasteful consumption. The largest industrial players are openly playing prisoner's dilemma, refusing to co-operate lest competitors take advantage.

The world may gain advantage from co-operative scientific

research, but there is not necessarily any advantage to the individual who does it. Indeed, such individuals may well have to sacrifice the k's, megs and gigs that could have been earned by applying their talents in the 'real world'. Nor will the benefit of fundamental research necessarily go to the country which funds it. Rather, as the great mathematician David Hilbert said, in the aftermath of the First World War, 'for mathematics, the whole world is a single country'. That is the climate of reason, and the climate of the planet also needs genuinely united nations rather than competing warlords. A splendid Victorian hymn starts with Greenland's icy mountains and Afric's coral strand. When in pessimistic mood, I suspect that both may disappear before any politics can fit the science.

Page three news

Politicians speak of being judged by history, although historians aren't judges and, when they do make judgments, disagree as much as anyone else. Posterity is a forgetful and feckless beneficiary, and there may not be one to thank the present for much. Posterity has famously never done anything in return: this is a sad aspect to the asymmetry of time. But mathematics is like time travelling, and can make some names live far outside their physical lives. A hundred years ago in old Europe, Raymond Poincaré was the leading statesman of France, and her president in the First World War. But it was his cousin Henri Poincaré who makes news today, though news of a kind that the world doesn't find easy to absorb.

In 1904 Poincaré had published a conjecture: that *every simply connected closed 3-manifold is homeomorphic to the 3-sphere*. As Alan Turing said, 'Conjectures are of great importance since they suggest useful lines of research.' And so it did for a century, stimulating new understanding of many-dimensional spaces, but remaining an unanswered question. The significance of Three is that this three-dimensional problem is actually harder than its analogue in any other number of dimensions. In 2000 it was adopted as a Millennium problem by the Clay Institute, offering a million-dollar prize. This question now appears to have been settled by a sequence of difficult ideas, culminating in the work of an individualist Russian, Grigory Perelman. This has shown that Poincaré's conjecture is correct. The prize money made it news, as it was intended to, but explaining the nature of the achievement brought accessibility to a crisis. The world's press rose to the occasion pretty well, though there was spluttering hilarity on the main BBC news programme when Poincaré's conjecture was read out.

What does it mean? A *manifold* is a space created by joining together patches, rather as fields of view are welded into the whole sphere of sight. *Closed* means having no edge. *Simple-connectedness* is a criterion of having no hole: it means that any circle drawn in the manifold can be continuously contracted to a point. A spherical surface is simply connected. The surface of a teacup, in contrast, has a hole: you can draw a circle on the handle which cannot be contracted. The property of having no hole is also shared by any stretched or squashed sphere. This is the significance of the word *homeomorphic*: it gives an exact definition of stretching and squashing. The mathematical theory of *topology* makes all these concepts precise, and it can be asserted that surfaces homeomorphic to the sphere are the only two-dimensional manifolds that have neither edge nor hole. Poincaré conjectured that the analogous statement was true in three dimensions.

We have already met an example of a closed three-dimensional manifold: it is the space of rotations described in Chapter 2. The shoe-string experiment relates to Poincaré's conjecture rather neatly. The fact that a *single* rotation leads to a twisting up, is equivalent to the fact that the space of rotations is not simply connected. That is consistent with the fact that the space of rotations is *not* the 3-sphere, but only half of it.

I was slightly surprised to hear a media-savvy mathematician explain that the question concerned what shape the universe could take, presumably thinking of its three spatial dimensions. This understates the scope of Poincaré's problem. The manifold should be thought of as a completely general conception of what three parameters could encompass, as a space of possibilities. As a physical application of the kind of research Poincaré inspired, the inner space of quantum particles and forces is more pertinent.

Poincaré's work echoes to the honour of France and her Third Republic, but was not done for profit. The same applies to another famous problem Poincaré attacked, also featuring the number Three. This is the *three-body problem*.

Climates and comets

The bodies that Poincaré was concerned with were of the heavenly kind. The classical *two-body problem* arises for double stars, attracting each other according to Newton's Law of gravitation. At first sight each star is characterised by six numbers which say where it is, and how it is moving. That would seem to make the problem one of a twelve-dimensional space of possibilities. But the symmetries of the problem reduce this to a total dimension of just two. In this two-dimensional space each possible story of the stars is traced out as a one-dimensional track. In fact, by making use of the idea of a space of possibilities, Newton's laws become elegant geometrical rules dictating where these tracks must go. The whole ensemble of possible stories can be visualised as a system of tracks neatly covering a surface. Stability is a natural feature in this two-dimensional setting.

But for *three* or more bodies the situation is entirely different, with six extra dimensions of possibilities for each body. In two dimensions, there is not much scope for what tracks can do: it is rather like brushing short hair on a two-dimensional head. In three or more dimensions, beehives and Mohicans give a hint of how Newton's laws can make them wind round in far more complex ways, more strangely than punk dreams. They may give the appearance of stability, but only as a temporary approximation.

Poincaré could not pursue detailed calculation of these complicated effects, but gained enormous insight from the general principle of treating the behaviour of a physical system as a track in a possibility-space of many dimensions. In the 1950s, when computers first became available, people tried to model the atmosphere for weather prediction. They effectively rediscovered what Poincaré had pointed out, and since then 'chaos' has been the subject of an enormous amount of new research.

Does chaos render climate prediction a forlorn hope? Are scientists reduced to waving their hands over the 'butterfly effect', unable to do better than fortune-tellers or astrologers? By the 1970s, as chaos became popular, people were indeed saying that any such complex system was hopelessly unpredictable. Now the prevailing view is that by trying out predictions based on many slightly differing scenarios, and leaving answers in terms of probabilities, it is possible to assert reliable statements about climate change. Another important factor is that the *average* temperatures involved in climate do not depend on the detailed butterfly effects of daily weather patterns. Even so, the work stemming from Poincaré's discovery shows that a system may shift quite suddenly to a new global pattern, with no obligation to return to its original state. If the human species wipes itself out, then after a few thousand years CO_2 might return to pre-industrial levels, but with the Arctic ice-free and the deep ocean currents like the Gulf Stream permanently changed.

Newton, applying his theory to the two-body problem, could explain the basic fact of the elliptical planetary orbits. But finer detail, including the effect of the planets on each other, has required much more modern insights, and has stimulated further centuries of mathematical development. Poincaré's analysis arose from the very difficult problem of predicting the exact motion of the Moon. Asteroids show chaotic behaviour. Enormous computer calculations are now employed, and it is difficult to recover the excitement of the seventeenth century, when the solar system became the test-bed for mathematical prediction. The periodic return of Halley's comet, whose successful prediction was one great early achievement, is a reminder of that original drama of the skies.

In *The Decline and Fall of the Roman Empire*, Edward Gibbon wrote an unusual passage reflecting, from the vantage point of the eighteenth century, on a comet visible in the sixth. His narrative suddenly adopted the out-of-time standpoint of science, reflecting on the return of the comet at 575-year intervals throughout human history, the latest having occurred in his own 'enlightened age'. Gibbon explained that the 'mathematical science of Bernoulli, Newton, and Halley investigated the laws of its revolutions', and that the next return was predicted for 2255. Gibbon imagined 'astronomers of some future capital in the Siberian or American wilderness' verifying the prediction.

The passage is unfortunately based on a misidentification of the comet of 1680 with earlier comets, with the 575-year period based on jumping to wrong conclusions, and not at all on what Newton and Halley had found out. In contrast, Gibbon's political prediction (so strikingly eclipsing his own Europe, then approaching its zenith) reads extraordinarily well. It would have been almost spot-on for the 1960s. It is now dated, not only politically, but because astronomers are no longer mere observers, but can design space probes to intercept comets. His bold vision is a salutary reminder that climate predictions which look ahead to 2100 are still short-sighted. That year is nearer to us than Poincaré himself. A strand of mathematical thought may easily take a century to unwind, as Poincaré's conjecture shows.

One of the most fascinating questions about climate is to what extent past human history has already been profoundly affected by climate change – a subject not treated by Gibbon – and to what extent it has, through the invention of agriculture, already *caused* climate change. The interrelation of humanity and the atmosphere lies somewhere in between the simplicity of comets, and the complexity of history, the subject that Gibbon described as the crimes and follies of mankind.

Heart of the matter

For the really, really, really pessimistic, Three brings little cheer. It is the number of jealousy, and the world of the arts supplies copious three-body problem pages. Three is the number of dividing and ruling. Three is the number of fighting an ally over how best to oppose an enemy. Three is the number of triangulation, which seems to mean doing what your enemies want, while leaving your friends with nowhere else to go.

That masterpiece of the worst-case scenario, Orwell's *Nineteen Eighty-Four*, is striking in its appeal to the language of numbers: it has a number for its title, and famously begins with a thirteen. (The brilliant 'more equal than others' of *Animal Farm* had already shown Orwell's penchant for a mathematical imagery of truth.) Three-ness enters deeply into its plot, which is a prisoner's dilemma, though the depth of pessimism is such that there is no dilemma. Winston Smith knows he will betray Julia and vice versa, and that both will co-operate with Oceania. Orwell gave a sophisticated picture of superpower politics, as a three-person game with unpredictable shifts, but hidden co-operation.

His analysis was based on the sudden and totally unprincipled pact between Germany and the USSR in 1939, reversed equally suddenly in June 1941, with alliances again shifting in 1945 when the American nuclear bomb developed for use against Germany became one which was, in effect, aimed at the USSR. Quantum mechanics is never far away. Robert Oppenheimer, the mathematical physicist in charge of developing the atomic bomb in 1945, was the same Oppenheimer accused and discredited as a 'security risk' in 1954. In the Cold War context, technical arguments about the hydrogen bomb were inextricably involved with individual and political conflicts.

The nuclear test of July 1945 was code-named Trinity. No one knew then that there was a deeper level to the nucleus than was touched by that explosion, and that the number Three would turn out to be vital to it. In contrast to the growth of nuclear weapons from that point, now far bigger than those of 1945 and still proliferating, the post-war world also resumed the amazing progress of global co-operation on inner nuclear space, peaceful, beautiful, creative and hugely surprising – and almost totally incomprehensible to citizens of the planet.

At the heart of matter

What keeps the nucleus together? The positive electric charges of the protons cause them to repel each other vigorously. So there must exist some other attractive force between the protons and the neutrons in the nucleus, to counteract that repulsion. This force was called the 'nuclear force' and by the 1960s it could be given a working description. Particles called 'mesons' were identified as its carriers. But as accelerators became powerful enough for collisions to penetrate single protons and neutrons, something quite unexpected was found.

There was evidence of *three* constituent parts inside each of these nuclear particles. The name 'quark' was invented by physicist Murray Gell-Mann for obscure reasons – the line 'Three quarks for Muster Mark!' from Joyce's *Finnegans Wake* has often been quoted. The situation was about as clear as Joyce's prose, and just to add to the confusion, there were two different senses in which 'three quarks' were involved. There seemed to be three quarks inside a nuclear particle, but there were also three kinds of quark – in modern terms, three 'flavours' of quark were then known. The proton had two 'up-quarks' and a 'down-quark' and the neutron had two 'down-quarks' and an 'up-quark'. The third flavour was the 'strange-quark', and this made excellent new sense of a zoo of unstable 'strange' particles that had been observed since the 1950s.

But the presence of three quarks in a nuclear particle made no sense at all, and for years there was much scepticism over whether they could be genuine entities. Quarks would necessarily be fermions, and the 'exclusion principle' dictated that there was a limit to the number that could be inside the nuclear particle. There would be room for two of the same kind, one of each spin, but not three. Yet one particular nuclear particle (the Δ^{++}) seemed to consist of three up-quarks. This was apparently impossible.

The way out of this was to propose that the quarks differed in some other, hitherto invisible quality. This unknown quality was given the name 'colour': nothing to do with light, but simply a reference to the three-dimensionality of human colour space. As a metaphor, the three quarks in a nuclear particle could be 'red', 'blue' and 'green', and then the overall particle would be an uncoloured 'white'. This step into the unknown turned out to explain far more than the original puzzle.

With electric charge, the duality of positive and negative lies within a single dimension: charges add up and cancel like numbers on a line. The three dimensions of quark colour give a naturally analogous theory, in which the 'colour charge' generates a new three-dimensional 'colour force'. Although the mathematical principle was very clear and simple, it took over 30 years to see the force of its predictions and check them against observations. One reason was that the resulting force was completely different from what anyone expected.

Electric force becomes stronger and stronger as two charges approach. Colour force is the other way round. It acts like a spring, becoming stronger as two colour charges separate, and diminishing to nothing as they approach. When this was understood, it explained why no quarks were ever seen alone: they were always 'confined' by colour springs. Strictly speaking, no complete proof of these properties of the colour force has yet been given, and that is the subject of another Millennium Prize problem. However, enormous computer calculations, especially since 2000, have made it all consistently credible. The subject is in the tradition of Oppenheimer, and the long calculations necessary for the atomic bomb, but also in debt to Poincaré, not so much because concerned with three bodies, but because it involves deep understanding of many-dimensional spaces. It brings together the most abstract mathematics and the most sophisticated high-energy experiments. Fortunately, mathematics comes particularly cheap.

The older observations of the nuclear force have gradually fallen into place. Quarks have their anti-particles, and the mesons turned out to be identifiable as quark-antiquark pairs. The large masses of the nuclear particles (like the proton, 1836 times heavier than the electron) can be explained in terms of more primitive quark parameters. What was originally called the nuclear force, between protons and nuetrons, can now be seen as a second-order effect of the colour-force acting on the constituent quarks. That second-order effect is still strong enough to bind about 240 nucleons into the nucleus of a large atom. As a rough analogy, an atom has no overall electric charge, but it still interacts via electromagnetic forces with other atoms. Indeed nuclear physics is now in a somewhat similar state to that of chemistry 80 years ago, when its atomic numbers and weights could at last be explained in terms of smaller and simpler things. Foremost amongst those smaller and simpler things is the nuclear number Three.

The 1960s picture of three different flavours of quark – up, down and strange – has turned out to be incomplete. There are (apparently) just six, with the remaining flavours now bearing the vaguely risqué names of 'charm', 'top' and 'bottom' ('truth' and 'beauty' were wittier suggestions for the last two, but seem to have lost out). Yet the number Three is still central in this picture of flavour, because the six are arranged in three 'generations' of pairs. This fundamental Three-ness is not understood at all, except in that there are reasons why there should be the same number of generations as are found in the non-nuclear (electron-like) particles. The weak force is now seen as penetrating inside the nucleon and changing the flavour of its quarks. It is still the weirdest force, muddling up the three generations of quarks and the three generations of electrons. Although based on the numbers One, Two, Three, there is a complexity in these fundamental forces that still eludes simple explanation.

Quantum chromodynamics, as the colour force is called, has ended the tyranny of Two. Polarities, lines of force, attraction of opposites, give way to a new troilistic music of unjealous threesomes. This is a surprise; in fact the whole thing is a surprise. In the 1960s it would have seemed far too much to hope that the interior of the proton could be explained and calculated accurately by a three-dimensional analogue of electromagnetism.

Robert Oppenheimer, trapped in the either/or logic of the Cold War, died while this nuclear revelation was in its early and tentative stage. As so often, it took too long for the timescale of individual human life. But Oppenheimer had made another, less famous contribution. On that three-based date, 1 September 1939, as Germany attacked Poland, he and his collaborator published a paper: *On continued gravitational contraction*. It was one of the first to take seriously the possibility that a star might 'close itself off from any communication with a distant observer; only its gravitational field persists'. Too late for him, *black holes* were only taken seriously 30 years later. They demanded a better understanding of –

4

It's a Square World

Everyone knows that two and two make four. Not everyone sees that two and two make four in several different ways:

$$2 + 2 = 4, 2 \times 2 = 4, 2^2 = 4.$$

In short, four is the square of two. Musically, a beat of four, in two twos, gives common time. This rhythm is the first to have an internal structure, a double duality, a micro-drama to every bar. Hidden under the surface, this beat will be found in the numbers.

Squares are special. The periodic table of the chemical elements shows that the elements are built up from certain magic numbers: 2, 8, 18, 32. These numbers all come from squares: $2 \times 1, 2 \times 4, 2 \times 9, 2 \times 16$, although the usual layout of the table fails to make this plain. The factor of two is the factor of spin already described in Chapter 2. Quantum mechanics is also responsible for the pattern of squares. They count the number of possible states for the electrons with the greater energies required for larger atoms.

So the chemistry of elements follows from the properties of squares. Carbon, in particular, being element 6, is just halfway along the sequence of eight that forms the second tier of elements, because $6 = 2 \times 1^2 + \frac{1}{2}(2 \times 2^2)$. This means its outer shell of electrons is just half-complete, and it can share *four* electrons. Oxygen, element 8, has just *two* places free. Because $2 \times 2 = 4$, they bond in carbon dioxide.

Go fourth

An $n \times n$ Latin square consists of n copies of the numbers 1 to n, placed in n rows and n columns such that each row and each column contains all the numbers just once. Does this puzzle sound familiar? A Sudoku solution is a 9×9 Latin square, with an extra condition on the 3×3 subsquares.

Latin squares can be used for devising a duty roster for n dirty jobs in a houseshare of n people. Or, for comparing the effect of n drugs on n animals in some doubtless vital trial. Who said mathematics wasn't useful? Similar ideas lead to the error-correcting codes which make it possible for computers to communicate reliably. Or, for that matter, to football leagues, speed-dating nights, and the plot lines for *Desperate Housewives*.

Latin squares are also connected with another kind

of square, presenting an even more complicated challenge: this is the multiplication table beloved of the Department of Education.

For some reason, much emphasis is placed on children learning it in a one-dimensional sequential order, when in fact the whole point of a table is to *see* it in two dimensions. Rote-learning misses the vital point that a x b = b x a, which you can see at a glance in a table. This symmetry is equivalent to the fact that you can reflect the picture in the NW-SE diagonal. Another symmetry comes to light if you ignore everything but the *last* figure of each entry. These last figures make the pattern:

1	2	3	4	5	6	7	8	9
2	4	6	8	0	2	4	6	8
3	6	9	2	5	8	1	4	7
4	8	2	6	0	4	8	2	6
5	0	5	0	5	0	5	0	5
6	2	8	4	0	6	2	8	4
7	4	1	8	5	2	9	6	3
8	6	4	2	0	8	6	4	2
9	8	7	6	5	4	3	2	1

showing another symmetry on the NE-SW diagonal. This second symmetry shows up in the way the nine-times table runs 9, 18, 27, 36... with the last figure always decreasing.

Together, they give a *four-fold* symmetry of multiplication.

The pattern is not a Latin square, but you can find patterns like Latin squares inside it:

1	3	7	9
3	9	1	7
7	1	9	3
9	7	3	1

4	8	2	6
8	6	4	2
2	4	6	8
6	2	8	4

You could study these patterns of the multiplication table in any number base. There is always a four-fold symmetry. Every number base has a different pattern, which tells a story about how it breaks down into factors. In a prime number base, the whole table is a Latin square.

Gauss found a subtle pattern in the numbers appearing in the *diagonal* of the square, which contains the last figures of the square numbers. In base 10, these are (1, 4, 5, 6, 9). In base 8, as you can see from Chapter 1, they are (0, 1, 4). Each base has its own selection of diagonal numbers, which at first sight look completely random. It took Gauss until he was nineteen before he found the key to the pattern, which we will come to a little later. This is why the multiplication table is a rather difficult subject for eight-year-olds.

It's a sin

At school you're told you must never divide by zero. So

young people with any self-respect will do it to see what happens. As usual the Pet Shop Boys are right on target with 'Two divided by zero': there is a problem. The problem arises because division is the inverse of multiplication. But a multiplication by zero cannot be undone: there is a broken symmetry.

If you are a drug dealer, you are probably good at dividing things into fractions. If two customers turn up, you know how to divide your stock in two. If you have no supply and two customers, that's also easy, if unprofitable. If you do have a supply but there are no customers, it's impossible to get rid of it. So 1/0 has no possible meaning. If you have no supply and no customers, then you can allocate a million deals to them all and it makes no difference. So 0/0 can be anything you like.

Constance Reid dealt carefully with 1/0 and 0/0 (though not using quite the same practical illustration) in Chapter 0 of her book. But she omitted the very question that makes 0/0 so interesting: it leads to the differential calculus, which describes how continuous things can change. Newton was sinfully dividing zero by zero all the time, to create this modern idea. He was ticked off by philosophers, but his calculus still worked wonders. It was not until well into the nineteenth century that this kind of 0/0 was given a satisfactory explanation.

In this book, as in hers, such questions of change and continuity must remain off-topic. But we do need a few basic terms to extend the discussion of adding and multiplying and taking powers.

Unearthly powers

Squaring a number – multiplying it by itself – gives a first interesting example of a *function*. That technical word 'function' is equivalent to a *graph*, where if the value of one thing is given, the value of another can be read off. A graph turns formulas for functions into pictures which suit the two-dimensional eye.

Taking the cube of a number is also a function, and so is the *n*th power – multiplying a number by itself *n* times. There is a pattern to powers: $2^m \times 2^n = 2^{m+n}$, and to be consistent with this, the power 0 is defined so that $2^0 = 1$. It is also consistently true that $(2^a)^b = 2^{a \times b}$. Halving is the inverse of doubling, and it is consistent to write:.

$$2^{-1} = 1/2, \; 2^{-n} = 1/2^n.$$

Then 10^{-3} metres is a thousandth of a metre, a millimetre; 10^{-9} metres is a billionth of a metre, a nanometre.

The next step is to go from squares to square roots, which means undoing a square. A graph gives a new way of looking at the duality of doing and undoing. It turns it into a reflection in the diagonal line:

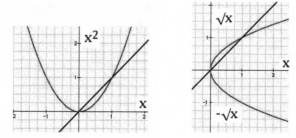

The number 4 has a square root: 2. But there is no integer which when squared gives 2. If asked for $\sqrt{2}$, a calculator will give an answer like 1.41421356, opening up a problem which we will soon put under a microscope. Another way of thinking of $\sqrt{2}$ is this: a multiplication by $\sqrt{2}$ is just halfway to doubling. So $\sqrt{2}$ can be written consistently as $2^{1/2}$.

GENTLE: If your shares go up 41.42%, and then increase again by 41.42%, they will (almost exactly) double.

Likewise the cube root can be seen as undoing the cube function, and the cube root of 2 written as $2^{1/3}$.

This notation naturally extends so that $2^{2/3}$ means the square of cube root, $2^{m/n}$ the mth power of the nth root. The fuller picture involves the fact that there are two numbers whose square is 2, because $(-\sqrt{2})^2$ is also 2. This can also be read off from the graph. These complications fall into place when the theory is extended to complex numbers, and the graphs become surfaces in $2 + 2 = 4$ dimensions!

A web of logs

The logarithm is another example of an inverse operation. It has a subtle relationship to roots. If b is the nth power of a, then the inverse statement is that a is the nth root of b: it answers the question, what number when raised to the nth power gives b? But there is another inverse, which arises from the question, to what power must a be raised to obtain b? The number n, which has this property, is called the *logarithm to base a* of b, written $\log_a(b)$.

The statement that 10^3 = 1000 can be put in reverse by saying that 3 is the logarithm of 1000 to base 10. Likewise, because 2^{10} = 1024, 10 is the logarithm of 1024 to base 2.

Logarithms are natural in describing the scale or 'order of magnitude' of a number. Roughly speaking, the logarithm to base 10 of a number measures its length when written in base 10. If you have a kiss-and-tell story to sell, and negotiate a six-figure sum, that means that the logarithm of your earnings, to base 10, lies (probably an appropriate verb) between 5 and 6.

Logarithmic charts are useful in scanning the stock market for speculative purposes. Eurotunnel shares, for instance, once went up from 500p to 1000p; now they are worth pennies. But if they now rise from 20p to 40p, that is just as good an opportunity for making a profit; logarithmic charts show this best by portraying a doubling as the same distance on the chart no matter what the price is. (If you invested your life savings in Eurotunnel, however, logarithms won't help you dig your way out.)

Logarithms also come into the 'magnitudes' of stars and earthquakes. (No one seems yet to have invented a scale for celebrities.) The Richter scale is logarithmic: for every two points on the scale there is a thousand-fold increase in energy. Size 12 would destroy the Earth. If the Earth moves for you and your partner, on the other hand, the energy involved is a shock of about minus 2 on the Richter scale. A pin dropping, or a passionless peck, has the impact of a Richter scale shock of magnitude about minus 4.

Decibels and speaker gain are logarithmic. So indeed are

the music notes themselves; the names ABC…, the 'third' and 'fifth' and so on, the blobs on staves, the numberings of guitar chords, are all roughly logarithms of frequency. The idea that *multiplying* a frequency by some ratio is equivalent to *adding* notes in a scale, is just the idea of the logarithm. Writing a careers advice column in the *Independent*, the singer Billy Bragg probably spoke for many when he said he had rejected science at school because 'I didn't see the point of logarithms'. But he did.

MODERATE: If a virus doubles its numbers every day, how many days does it take to increase by a million? If $2^2 = 4$, what is the logarithm of 4 to base 2? What is the logarithm of 2 to base 4?

TRICKY: Why is $\log_{10} 2$ very close to 0.3?

Constance Reid wisely left alone the even naughtier question of what 0^0 means. According to a school maths website, '$x^0 = 1$ (unless $x = 0$, then $x^0 = 0$)'. The 'rule' that $0^0 = 0$ has perhaps been arrived at like this: you can say that the square root of 0 is 0, and likewise $0^{1/n} = 0$; so $0^{m/n} = 0$, so $0^x = 0$ for all x, so $0^0 = 0$. But you could equally well argue that $x^0 = 1$ for all x, and that it cannot suddenly jump when x becomes 0. This is a good example of the difficulties in the concepts of continuity and limits that had to be sorted out in the late nineteenth century for the calculus to make sense. The full story involves studying powers and logarithms for complex numbers, but it does not produce an answer for 0^0, which is as ill-defined as $0/0$.

Calling the tune

Powers and logarithms provide the Western solution to the problem of musical scales. The solution is a fudge, a creative accountancy called equal temperament, but a fiddle commended by Bach and Mozart commands respect. The secret of the equal-temperament musical scale lies in fractional powers. If twelve equal semitones make an octave, a doubling of frequency, then the semitone ratio must be given by the twelfth root of two, which is about 1.059463. If the note A has a frequency of 440 cycles per second, then B♭ is at 440 x $2^{1/12}$, B at 440 x $2^{2/12}$ and so on.

Equal temperament is another example of synthesis overcoming a deadlocked either/or. The price paid is that not a single interval, except the octave, is correctly in tune. Each one is a just-tolerable approximation. But harmony is thereby established on Western terms. Is this a familiar puzzle?

The major third is approximated by four equal-tempered semitones, giving a ratio equal to $2^{1/3}$, the cube root of two. This is very close to 1.26, and so less than 1% sharp of the pure harmonic third given by the ratio 5/4 = 1.25. Equivalently, $\log_2(5/4)$ is near 1/3. Put together with the eight-semitone interval, this gives the augmented triad with its three-way symmetry. Richard Strauss used this triad as the fugue subject in his 1896 tone-poem *Also Sprach Zarathustra*, and thereby expressed the relentless progress of cold soulless science etc. etc. etc.

The square root of two, equivalent to six equal-tempered semitones, is the interval remotest from the simple harmonics.

The interval is tame as a pussy-cat when sitting in the comfy lap of the dominant seventh chord, but standing alone it is what eighteenth-century writers called the *diabolus in musica*, certainly the devil to sing.

Four last things

An interval of *three* equal-tempered semitones corresponds to $2^{1/4}$, the *fourth* root of two, about 1.189. This is to be compared with a pure minor third, with harmonic ratio $6/5 = 1.2$. It differs by less than 1%. A chord of four such intervals is called the diminished seventh, exploited in classical-classical music for its melodramatic misery and unique four-way symmetry between four keys, which allows for entering it from one key and leaving in another. Mozart used this symmetry in an amazing passage, the end of the 'Confutatis maledictis' of his Requiem, emphasised by four dramatic bass leaps of the *diabolus in musica*. Probably one of the last passages he wrote, it is as free in its ever-shifting tonality as the *Vier letzte Lieder (Four last songs)* of Richard Strauss 150 years later.

Indian and Arabic musical scales have far more subtle and varied tuning solutions for melody, but it is western polyphonic harmony, like African rhythm, that has globalised, through the twelfth roots of 2. It is possible that the Arabic scale went into the Islamic African culture that, enslaved in the Americas, gave rise to blue notes. If so, the call of the muezzin, which had little impact on classical Europe, has morphed into one of America's major cultural exports.

Meanwhile European art music of the early twentieth century got carried away by twelfth roots, abandoning the base in harmonics, and drove itself into a tiny niche. But the avant-garde use of electronically generated sound in the 1960s made a great hit and rapidly found its way into popular music.

Electronic synthesis extends all options, of tonality, melody and harmony, rhythm and timbre. On the electronic scale of nanoseconds, a single sound wave takes a long, languorous age and can be constructed at leisure. Particularly popular is the construction of low sounds which, at the borderline of sonic and tactile, in pre-electronic days, needed a huge investment in vast pipes like church organs. Now you can do this in your house: hence dance music. Music technology creates sound spaces of virtually unlimited numbers of dimensions. The banks of controls and switches and MIDI channels of synthetic music are, in their prolific generosity, a picture of the huge numbers of parameters available.

The space of possibilities for sound is based on the theory of waveforms. Joseph Fourier, who founded this theory in the 1820s, developed it not for music but for the study of heat, making strong use of the duality of growth and oscillation. He also applied his theory of heat to give the first theory of the Earth's temperature – though in those days, before infra-red radiation was understood, it was only guesswork. His work gave the first suggestion of an atmospheric 'greenhouse effect'. This illustrates the unity and unexpectedness of mathematical exploration. Science needs synthesis as well as music.

Creating sound with electronic synthesisers is not so different in principle from fooling the eye with pigments. But in the class war of the arts, the former counts for little, and the latter counts for much. A philosopher rashly claimed the Pet Shop Boys, leading expositors of synthesised sound, had made only a 'minimal contribution' to their music, and that 'sound engineers' had done it all. He underestimated these artists. They sued. He paid up. This kind of judgment overlooks the relationship between artistic composition and the mathematics of sound. Mozart also needed instrument makers who could engineer twelfth roots.

Writer's block

The roots of 2 are also vital for literary composition. When I was first asked to write a column about numbers for the *Observer*, I faced a blank sheet of paper and came up with – a blank sheet of paper. More precisely, a sheet of A4 paper, as is now used in all the un-American world: it has been a German standard since 1922. The defining property of the series A0, A1, A2, A3, A4, A5... is that each size can be obtained by cutting the larger one in half – efficient both for manufacture and for photocopying – and that they all have the same shape.

If you wanted to create such a shape, without using a calculator, how would you do it? First, suppose you are given a square of paper, and are asked to find a square with just half its area. You can do it all by folding: first find the mid-points of the sides by folding and then fold along diagonals:

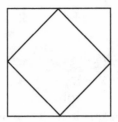

Then the inner square clearly has the required property. What is the length of its side, compared with the sides of the original square? If the original square had a side of 2 units, then the inner square must have an area half of 4, which is 2, and so its side must be $\sqrt{2}$.

The A4 rectangle shape then comes from using the length of the diagonal as the shorter side and the side of the square as the longer side. Or, equally well, the length of the diagonal as the longer side and half the side of the square as the shorter side. You can check this by laying three A4 pages together like this:

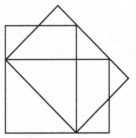

An additional property is that an A0 sheet has an area of 1 square metre. This leads to a complete formula for the size of An paper: Long side: $2^{1/4 - n/2}$ m, short side $2^{-1/4 - n/2}$ m, area 2^{-n} m^2.

The B paper sizes are such that Bn is A($n - 1/2$). Organ pipes with lengths taken from A and B paper sides will play a perfect chord of the diminished seventh. This is an example of what Wagner saw as the union of the arts.

Roots and trees

But what, exactly, is the square root of two? So far I have avoided this question by saying it is 'about' some value. A calculator will give a value like 1.41421356, depending on its precision. But however many decimal places the calculator offers, what it says will never be exact. We know this because if it were exactly the decimal 1.414, for instance, it would be an exact fraction 1414/1000. But it cannot be any such ratio of natural numbers. The ancient Greek mathematicians knew this and gave a proof without the benefit of A4 paper – or decimals, for that matter. It is a famous model of logical argument involving a proof by contradiction, quoted by Hardy in his 1940 *Apology* as a perfect example of its kind.

The argument, like much in this chapter, depends on the fact that $2 \times 2 = 4$. It goes like this. Square numbers are either even or odd. If they are even then they are squares of an even number, and that means they are divisible not just by 2 but by 4. Now suppose if possible $\sqrt{2}$ is equal to an exact fraction m/n, where m and n are not both even numbers. (If they are both even, divide both by 2 and carry on until one of them is not.) Then $2 = m^2/n^2$, so $2n^2 = m^2$, so m is even, so m^2 is divisible by 4, so n^2 is divisible by 2, so n is even. This is a contradiction. So no such numbers m, n can exist.

This shows that *irrational* numbers – numbers which are not ratios of integers – are inescapable in any discussion of *space*. It also points to wider issues, which took millennia to sort out. Is geometry a purely logical theory of the continuum of numbers, or a physical theory of space? Even Gauss does not seem to have been clear, and these concepts only gradually separated in the nineteenth century. By 1900, they were both ready for different and radical developments. The numbers, separated from physical embodiment, were ready for the logical analysis of Russell and Gödel. Physical space, separated from Euclid's geometry, was ready for Einstein. Nowadays the conceptual distinction is clear, and yet in practice, mathematics interweaves as freely as ever. I shall let the number Four emphasise these interconnections.

A walk in the woods

How does the calculator work out its approximation? I would be surprised if it used the method for square roots that was taught in schools in pre-calculator days. This was based on the following idea: that by squaring, you find 1.4 too small, 1.5 too large, then 1.41 too small, but 1.42 too large, and so on. Although this can be usefully streamlined, it requires much laborious work. For a quicker method it is better to draw the following picture.

As you pass by a forestry plantation with a rectangular grid of trees, you may encounter avenues of light opening up in different directions. In the grid below, there are obviously vertical avenues and horizontal avenues, and there are also diagonal avenues (one up for one along) clearly visible. There are less obvious and narrower avenues corresponding to *every rational slope* — every slope that goes an integer *m* up for an integer *n* along. But if we take a line of slope $\sqrt{2}$, the argument above shows it does not have any such rational slope. However many avenues we explore, we shall never find it. If we draw such a line through one tree, at 0, it cannot pass through any other tree. Thus, on a clear day you can see forever, as long as you look in an irrational direction.

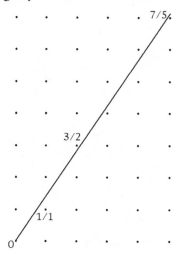

In this picture, good approximations to $\sqrt{2}$ appear as trees which are close to the line. The fractions 1/1, 3/2 and 7/5 on the grid are the first such nearby trees. The sequence continues: 1/1, 3/2, 7/5, 17/12, 41/29, 99/70, 239/169, 577/408, 1393/985...

Each tree gives an approximation which is about six times better than the last, as you can check with a calculator.

These are very efficient because there is a simple rule: each fraction gives the next by $(m, n) \rightarrow (m + 2 \times n, m + n)$. Even better, you can jump the queue, going immediately to a point twice as far along in the sequence, by using the rule $(m, n) \rightarrow (m^2 + 2 \times n^2, 2 \times m \times n)$. This does a little better than doubling the number of correct decimal places every time you apply it.

DIFFICULT: Show that $\sqrt{2}$ is approximately 886731088897/627013566048, correct to 22 decimal places.

The pairs of numbers (m, n) have a special property. There are no integers m, n that satisfy $m^2 = 2 \times n^2$, but these pairs come as near as possible to doing it: you will find with a calculator that m^2 differs from $2 \times n^2$ by 1 or -1 alternately. The problem of finding integer solutions to equations of this kind is called a *Diophantine* problem, after Diophantus, who wrote a text of problems and answers in the third century CE A Killer Sudoku puzzle is an example of a simple Diophantine problem which only involves addition.

FIENDISH: See why there are not any numbers m, n, with the square of m only differing by 1 from twice the square of n, except those appearing in the sequence above.

The twelve-tone scale uses the approximations $\log_2 3 = 19/12$, $\log_2 5 = 7/3$. One can look for alternative equal-temperament scales by using a similar walk-in-the-woods picture of approximating the irrational logarithms by nearby rationals. A scale with 19 equally separated notes is the next possible candidate. It does better than the twelve-note scale with the approximation $\log_2 5 = 44/19$. Scales of 31 and 53 are the next possibilities, and for microtonal enthusiasts, these observations are only a glimpse of the subtleties of tonality which can be explored.

Cutting corners

There is another way of looking at that folded square of A4 paper. It can be seen as an illustration of the theorem attributed to Pythagoras, which can be espressed as follows: given a rectangle of sides a and b, the length of its diagonal is $\sqrt{a^2 + b^2}$.

This is perhaps the simplest thing about squares that is not at all obvious to the eye. The most famous example uses the fact that $3^2 + 4^2 = 9 + 16 = 25 = 5^2$, and shows that the diagonal of a 3 x 4 rectangle is 5. Why is this true? From the modern point of view, in which geometry is a theory of continuous numbers, not a scientific account of physical space, Pythagoras's theorem is a *definition* of what is meant by distance. So the following picture is not a proof of it, but an illustration showing that it is consistent with what we expect of squares. In this case it won't suffice just to fold the square; it needs some surgery and rearrangement.

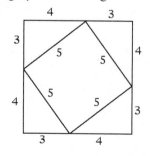

The area of the large square is $7^2 = 49$. The four triangles at the corners can be rearranged into two 3 x 4 rectangles, and so have area 24. This leaves 25 for the inner square, agreeing with its side being 5, as marked. The same argument will illustrate Pythagoras's Theorem in general, using the fact that $(a + b)^2 - 2 \times a \times b = a^2 + b^2$.

A classic Diophantine problem is that of finding all such triples of integers (a, b, c) such that $a^2 + b^2 = c^2$. The popular novel called *The Curious Incident of the Dog in the Night-time* gives an astonishingly long discussion of it. Here are some puzzles about other related patterns:

DIFFICULT: The following figures are right-angled triangles that have two sides nearly equal: (3, 4, 5), (20, 21, 29), (119, 120, 169), (696, 697, 985)... Find a rule for finding more of these. (Hint: this is closely related to the approximations for $\sqrt{2}$.)

MODERATE: Another famous story about the young Gauss is that when told at school to add up all the numbers from 1 to 100, he did it quickly by seeing a pattern. In Killer Sudoku you often need to know and use the sum of the numbers 1+2+3+4+5+6+7+8+9. One way of seeing it is to draw it as a *triangle*

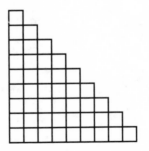

By putting two triangles like this together show that the sum is $\frac{1}{2}(9 \times 10) = 45$. Then show that the numbers from 1 to 100 add up to 5050.

FIENDISH: The triangular numbers are 1, 3, 6, 10, 15, 21, 28, 36, 45, 55... Of these, 1 and 36 are also square numbers. What other triangular numbers are also square? (Hint: this is also closely related to the approximations for $\sqrt{2}$.)

There is a Diophantine equation which is very simple to write down, but is famous for resisting centuries of attack. This is the equation of the great seventeenth-century mathematician, Pierre de Fermat: $a^n + b^n = c^n$. When $n = 2$ there are many solutions given by Pythagorean triples. But squares are special: there are *no* solutions for higher powers. That is, as Fermat asserted, there are no two cubes which sum to a cube, no two fourth powers that sum to a fourth power, and so on. It is famous mainly because Fermat apparently claimed a marvellous proof of this fact, but gave no indication of what it was except that it was too long for the margin of his edition of Diophantus. Fermat was not always right, and we would now say it was almost certainly only a conjecture, which like Poincaré's could only be justified very much later. It was proved by Andrew Wiles only in 1995, using advanced ideas in algebraic geometry of which the walk-in-the-woods gives only the faintest glimpse, and with which few mathematicians are familiar.

Another Diophantine problem leads in a different (and easier) direction. If you draw a rectangle with sides a and b which are integers, then $a^2 + b^2$ is usually not a square integer, so the diagonal will have an irrational length. We can draw $\sqrt{5}$ as the diagonal of a rectangle with sides 1 and 2. But $\sqrt{7}$ cannot be formed immediately in this way. Which square roots can arise as the diagonal of a rectangle, and which can't?

The march of primes

In other words, which numbers are like 2 and 5 in being sums of two squares? The answer is far from obvious: it lies in looking at the prime numbers and the music of common time, a beat of four.

Count off the numbers to a four-four rhythm: ONE two Three four | FIVE six Seven eight | NINE ten Eleven twelve | THIRTEEN... Then we have on-beat odd numbers and off-beat odd numbers. It turns out that their properties are sharply distinguished. Apart from 2, the prime numbers divide into on-beat (5, 13, 17, 29, 37...) and off-beat (3, 7, 11, 19, 23, 31...). There are infinitely many of each kind. The on-beat primes can *always* be written as the sum of two squares: $5 = 2^2 + 1^2$, $13 = 3^2 + 2^2$, $17 = 4^2 + 1^2$, $29 = 5^2 + 2^2$, $37 = 6^2 + 1^2$... The off-beat primes can *never* be so represented. This was discovered by two great mathematicians: announced by Fermat in 1640, but with a proof first published by Leonhard Euler, Gauss's predecessor, 100 years later.

Gauss's discovery about the diagonals of multiplication tables is a key to this deeper level to the numbers. It is useful first to think of those multiplication tables in another way. The cell that says 1 is the last figure of 3 x 7 also says that if you multiply any number ending in 3 by any number ending in 7, the result is a number ending in 1. The table is really about these *classes* of numbers, which are called *congruence classes*. Two integers are said to be congruent (modulo 10) if they end in the same figure. Equivalently, two integers are congruent (modulo 10) if they differ by a multiple of 10.

In the same way, all odd numbers are congruent to each other (modulo 2), and all even numbers are congruent to each other (modulo 2). The on-beat odd numbers are those that are congruent to 1 (modulo 4), and the off-beat odd numbers are congruent to 3 (modulo 4).

You are already familiar with congruence classes (modulo 7) from the days of the week. If the 1st of January is a Wednesday, what is the 30th of January? Because 30 is congruent to 2 (modulo 7), the question is the same as asking which day of the week the 2nd of January falls on, which is a Thursday.

Now we can state Gauss's discovery in terms of congruences. Take two primes p, q, not 2. Consider the multiplication table for numbers modulo p, and the table for numbers modulo q. If p and q are both off-beat, either p (modulo q) appears in the diagonal of the q-table, or q (modulo p) appears in the diagonal of the p-table, but not both. If one of p or q is on-beat, then either they both appear in each other's diagonals, or both are absent.

This is called the *Law of Quadratic Reciprocity*: the name reflects that it is an intricate duality of dualities. It shows that the primes form a hidden infinite Sudoku of consistency, all fitting in with each others' diagonals.

Factorising primes

Numbers which are built up by multiplying on-beat primes (and 2) also have the sum-of-two-squares property. For instance 5 and 29 are on-beat primes, and $145 = 29 \times 5$ also has the two-square property because $145 = 64 + 81 = 8^2 + 9^2$.

There is a way of seeing why this is so. When the number 2 makes an appearance, the complex numbers are quite likely to lurk behind the scenes. The two-square property means that on-beat primes can be thought of as *factorising* into complex integer pairs: thus $5 = (2, 1) \times (2, -1)$, $29 = (5, 2) \times (5, -2)$. From this point of view they are not actually primes at all!

In this light, we can look again at their product 145 as $(5, 2) \times (5, -2) \times (2, 1) \times (2, -1) = ((5, 2) \times (2, 1)) \times ((5, -2) \times (2, -1))$ by reordering the complex factors this can be expressed as$= (8, 9) \times (8, -9) = 8^2 + 9^2$.

In fact, this completely answers the question of which numbers are the sum of two squares: all numbers which break down entirely into on-beat primes. The underlying algebra can also be expressed by a formula that doesn't mention complex numbers:

$$(a^2 + b^2)(c^2 + d^2) = (a \times c - b \times d)^2 + (a \times d + b \times c)^2$$

At this point we go into fourth gear, and speed up. It turns out that *every* integer can be written as the sum of just *four* squares. There is also a corresponding formula:

$$(a^2 + b^2 + c^2 + d^2)\,(A^2 + B^2 + C^2 + D^2) =$$
$$(aA - bB - cC - dD)^2 + (aB + bA + cD - dC)^2 +$$
$$(aC + cA + dB - bD)^2 + (aD + dA + bC - cB)^2$$

To save space I have left out multiplication signs. The important feature is the pattern of plus and minus signs. It does not have the symmetry you might expect. Although it is a picture of a hidden double duality that lies inside the sequence of integers, it does not break the numbers into two pairs. It breaks that symmetry and exploits a deep but easier fact: $2 \times 2 = 1 + 3$.

Quaternions

Does a sum of four squares have some analogue of the factorisation of a sum of two squares into complex numbers? It does indeed: it can be thought of as splitting into *quadruplets* called *quaternions*. In Dublin, a plaque bearing the date of 16 October 1843 marks where the Irish mathematician William Hamilton 'in a flash of genius discovered the fundamental formula for quaternion multiplication'. But the quaternion multiplication that Hamilton discovered is not like the multiplication of numbers. The splitting of $2 \times 2 = 1 + 3$ reflects the fact that multiplying quaternions is like combining *three-dimensional rotations*.

This moment in Dublin marks a milestone in the process of writing new symbols for entities going beyond numbers. Hamilton's work was also of great importance for starting the description of physical systems in terms of spaces of possibilities, such as Poincaré later used, and which turned out to be vital to quantum mechanics.

This may seem abstruse, but quaternions have a direct application. They are needed for the huge industry of computer games or, if you prefer, animated training software like *Brain Surgery for Beginners*. Three-dimensional graphics need three-dimensional rotations, and quaternions code them efficiently. A more macho example would be the efficient handling of roll, pitch and yaw in aircraft or spacecraft. In the coding of rotations by quaternions, there is a pernickety point to observe. If Q is a quaternion coding a rotation, then –Q codes the same rotation. It is an inconvenient Two, but there is no getting rid of it: it is so.

That Two has a meaning. It is just the same as the Two-ness defining electron spin. That connection turns out to be highly significant, and so does that $2 \times 2 = 1 + 3$. A splitting of four into one and three is so familiar that it might go unnoticed. It is there implicitly in a statement about a date and place in Dublin: there is one dimension of time, and three dimensions of space. The next step in exploring the number Four requires some investigation of how time and space are connected.

Time and motion pictures

H. G. Wells's 1895 novel *The Time Machine* was not really about time at all, but about class society, with more than a hint of how arts-educated persons, parasitic on science and technology – the world of the subterranean Morlocks in his story – might one day get come-uppance. Nevertheless, Wells wrote a very striking introduction to the scientific concept of time as a dimension. 'There is no difference between Time and any of the three dimensions of Space except that our consciousness moves along it,' explains the fictional inventor, showing 'a portrait of a man at eight years old, another at fifteen, another at seventeen, another at twenty-three, and so on. All these are evidently sections, as it were, Three-Dimensional representations of his Four-Dimensioned being.' Ancestral trees are a simplified form of what Wells was suggesting, showing the time dimension as if it were a space dimension. Perhaps Wells was influenced by seeing the first reels revolving in a cinema, turning space into time.

But time is *not* the same as space. Something new and different has to come in, and that is what Einstein provided just ten years later. Pythagoras's Theorem is still the key.

First we must see how Pythagoras's Theorem extends from two to three space dimensions. As an example, suppose you are smuggling ancient Peruvian artefacts out of that country. You have a carved rod of length 13.5 units, and a rectangular box of length, width and height given by 12, 3 and 4 units. Obviously the rod will not fit in the box lengthwise, but can the other two dimensions be used to make room for it? We need to know the length of the complete diagonal of the box, i.e. the distance of a corner from the most remote opposite corner. Except when moving furniture through narrow doors, you may well not have had much cause to think about how such lengths are related. The answer comes from applying Pythagoras's theorem twice. First, see that the small faces of the box have a diagonal of $\sqrt{3^2 + 4^2} = 5$ units. Now, take a rectangular slice through the box based on this 5-unit diagonal, and the 12-unit length of the box. The diagonal of this slice, which is the complete diagonal of the box, is $\sqrt{5^2 + 12^2} = \sqrt{169} = 13$ units. Fortunately for Peru, the rod will not fit. In general, for a box of sides a, b, and c units, the complete diagonal has length $\sqrt{a^2 + b^2 + c^2}$ units.

Although for simplicity I cunningly chose lengths $(3, 4, 12)$ which ended up with an integer value for the complete diagonal, we are now thinking of lengths which may take any value. With this in mind, here are two useful formulas which show a pattern:

$x^2 + y^2 = 1$ is the equation of a unit circle: a one-dimensional space sitting in two dimensions. It accounts for all the points (x, y) that are at distance 1 from $(0, 0)$.

$x^2 + y^2 + z^2 = 1$ is the equation of a unit sphere: a two-dimensional space sitting in three dimensions. It accounts for all the points (x, y, z) that are at distance 1 from $(0, 0, 0)$.

The natural next step is to write:

$x^2 + y^2 + z^2 + w^2 = 1$ is the equation of a unit 3-sphere: a three-dimensional space sitting in four dimensions. It accounts for all the points (x, y, z, w) that are at distance 1 from $(0, 0, 0, 0)$.

This is the 3-sphere already mentioned in Chapters 2 and 3, but now given a definition. A practical use for this four-dimensional measure of distance comes in applying it to the 3-space of rotations parameterised by quaternions. A computer game could use it to roll something smoothly from one position to another.

A measure of distance is called a *metric*. There are many possible metrics on a space. If you are a supermarket chief you might, however unhealthily, want to give more weight to chocolate bars than to lettuce and muesli when choosing a metric on stock-control space. For the three-dimensional colour space, the question of a metric correctly measuring how close one colour seems to another would be a serious and difficult question in experimental psychology.

But what is the right metric for one physical dimension of time (t) and three of space (x, y, z)? Which metric is respected by the laws governing matter and forces? In H. G. Wells's picture, t is said to be just the same as x, y, z. If this were true then the measurement of distance should treat them alike, as $t^2 + x^2 + y^2 + z^2$. But time is plainly not just the same as space: *The Time Machine* is fiction. It turns out that there is a simple difference between fiction and fact. The right expression is: $t^2 - x^2 - y^2 - z^2$. That simple change of plus signs to minus signs is an enormous, radical step. By making that splitting of four into one and three, it expresses the Theory of *Relativity*.

The M-I-N-I-M-A-L contribution of the PSB has it right: '... light and shade, time and space.'

Uncommon time

The word 'relativity' has given rise to endless sermonising on twentieth-century moral relativism, the loss of absolute authority etc. etc. etc. It is a pity that this word became the keynote, because Einstein did not introduce the idea that 'everything is relative' at all. On the contrary, he showed how to give an absolute meaning to crucial physical ideas, which until then had seemed to depend on how you looked at them.

You may find it convenient to run 10km relative to a moving treadmill mat. Psychologically, the daytime television and air-conditioning of a gym may be very different from the breezes, birdsong and traffic hazards of a 10km road run. But the principle of the treadmill is still sound, because it is the principle of relativity. The physical effort you must make is essentially the same: it does not depend on what is going on around you or how you look at it. The problem for physicists before Einstein was that they could not see how to extend this basic idea to electricity, magnetism and light. These were beautifully unified with each other, but the unified theory was not consistent with Newton's laws.

Einstein achieved this extension by abandoning a Newtonian assumption which had always seemed natural, but was in fact unwarranted. If two people synchronise their watches, then move around in different ways, and then meet up again, will their watches agree? They will not. This is not a question about making accurate watches – we can assume them perfectly accurate. It is a question about the existence of a universal time that all perfect watches will tell. There is no such time co-ordinate.

'At this moment of time', anchorpersons say, to an audience 'out there'. To illustrate the difficulty of defining a universal time, something which defines 'this' moment of time at all places, imagine the breakfast show broadcast to the whole solar system. Far out there, near Jupiter, when astronauts catch Earth News, what are they to make of 'this' moment in time? They must allow for the time taken by the radio signal, which travelling at the speed of light takes several hours – but how many hours depends on their distance. The natural way to measure *that* is by sending yet more radio signals – just like police speed cameras making radar measurements of speed and distance, only on a larger scale. To establish time and distance requires a system of sending signals back and forth and doing calculations, and in no way is 'this moment of time' a primary, fundamental observation of reality. It is the signal itself that is fundamental; this is electromagnetic, and travels at the speed of light. The light-signal becomes the foundation of Einstein's theory.

Once this assumption of a universal time is dropped, the problem of 'relativity' disappears. The laws of electricity, magnetism and light, including the constant speed of light, are just the same whether you do experiments while running on a treadmill, or while running along a road. They become absolute laws, which do not depend on how you look at them.

Einstein's radicalism succeeded where others had shrunk back. Even Poincaré had failed to cut the Gordian knot. But it was not Einstein who gave the first statement of that simple metric in four dimensions which embodies and streamlines his ideas. This was done in 1907 by the mathematician Hermann Minkowski. His formula stated an absolute measurement of the separation between two events, $(0, 0, 0, 0)$ and (t, x, y, z). The rule is this: you take $t^2 - x^2 - y^2 - z^2$. If this is positive, its square root gives the absolute time difference between them. If it is negative then $\sqrt{-t^2 + x^2 + y^2 + z^2}$ gives the absolute spatial distance between them. If it is zero, then they are separated by a light ray, and are thus in immediate contact by a light signal. Minkowski said, 'Space by itself, and time by itself, are doomed to fade away into mere shadows, and only a kind of union of the two will preserve an independent reality.'

It is more usual to write the rule using $c^2t^2 - x^2 - y^2 - z^2$, where c is the speed of light. I have simplified it by assuming the use of units such that c is 1. If space is measured in feet and time in nanoseconds this is almost exactly right, and worth thinking about. Light is usually thought of as incredibly fast — a second from the Earth to the Moon. But for computer engineering, it is the slow speed of light that dominates everything. It forces the miniaturisation of computer chips, for which a nanosecond is nowadays a painfully long epoch. Computers now make a trillion operations in a second, so light crawls less than a millimetre when one is made, and this is why the chips must be tiny. For anything involving time co-ordination to this degree of accuracy, such as the Global Positioning System for navigation, engineering must absorb Minkowski's metric. The idea of 'now' lingers on poetically like the four-squareness, four corners, or four winds of a flat Earth. It is adequate for everyday purposes where light seems instantaneous, just as for many purposes it makes no difference that the Earth is round. But for anything serious, the correct metric must be used.

We are all doing time travel. For we all travel into the future. Yet some travel faster than others. Minkowski's metric implies that the closer you get to the speed of light, the faster you travel into the future. In principle you can get to the year 802,701 to see if Wells's dystopia will come true – but unlike his time traveller, you cannot then return to tell the tale to spellbound listeners. That is because time is not the same as space; it has only one dimension and it leaves no room to turn round and come back.

Unfortunately, fantastic accelerations (and so fantastic rocket science) are needed to achieve any significant effect. If you could run round at a circuit at 98% of the speed of light (like particles sent round cyclotrons) then after what seemed to you ten minutes, the rest of the gym would have experienced 50 minutes. To get 800,000 years ahead of the rest of the world, while only living a normal span of life, would require acceleration to 99.999999995% of the speed of light.

The so-called 'twin paradox' makes a puzzle out of this. The 'paradox' is supposed to be that if you are whizzing round relative to the gym, the gym is whizzing around relative to you, and that the argument can be applied in reverse. This misunderstanding comes from misguided talk about everything being relative. Acceleration is not a relative but an absolute concept. Far from being paradoxical, and far from heralding moral laxity, the lesson of Minkowski's metric is a very Victorian ethic: you get ahead only with hard work.

Dust to dust

By modern standards, the Earth's wobbly rotation is far too

irregular for telling the time. The second is defined by atomic clocks, and Earth-time has to be adjusted now and then by a 'leap second'. The atomic clock depends on electrons and protons being absolutely reliable clocks, thanks to constant rest-masses. That rest-mass is the m in that most famous formula with a square: $E = mc^2$. Here E stands for energy, c for the speed of light, so the formula asserts that there is an equivalence between mass and energy.

Nowadays, that equation is seen as just one aspect of adopting Minkowski's metric. It is not the whole story, but is valid for isolated, non-interacting point particles, often called 'dust'. If you are a desperate house-spouse, then you may like to think of the meaning of the equation as you do your socially atomised housework. Dust settles in houses at about $6mg/m^2$ every day; for a room of $10m^2$ this is 60mg in a day. $E = mc^2$ tells you that if this were converted to explosive energy it would yield about 5×10^{12} joules, equivalent to about 1000 tons of TNT. It is a curious fact that the explosion of a gram of TNT has the same energy as a calorie of the kind you count or don't count for diets. It also releases about the same energy as a bang measuring (–2) on the Richter scale. So your room collects 10^9 calories of dust, enough to feed 400,000 people. It would power a similar number of light-bulbs, or make a tremor measuring 4 on the Richter scale.

This is fun to work out but also as unlikely as a plot in (dare I say?) the third series of *Desperate Housewives*. Quantum mechanics – more correctly, its extension to *Quantum Field Theory* – places severe restrictions on how mass and energy can be traded. It is like a barter economy where you can only trade in a rather limited line of goods. This conversion to energy could only be complete in the unlikely event of half of the dust actually being anti-dust made of anti-matter, perhaps shed by visiting aliens dropping by.

Richard Feynman, the American physicist and mathematician in the Second World War nuclear bomb project, did much to create quantum field theory in the aftermath. He devised an exact accounting system for the barter of energy. Feynman diagrams, as they are called, could well be described as 'Feynman accounts'. They are book-keeping accounts, making sure that energy balances, but also stories, telling you what happens to particles and forces. With time running up the page, here are three typical Feynman accounts, examples of the very limited possibilities for trade.

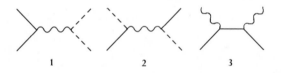

(1) Dust to dust, ashes to ashes, with an electromagnetic interaction between them (2) Dust to ashes, ashes to dust, with a weak interaction between them which can change the type of particle (3) Dust and anti-dust annihilate, leaving electromagnetic energy. In these accounts, 'dust' and 'ashes' stand for quarks or electrons of some flavour.

These basic meetings and partings build up into the more complicated orgies of interaction that occur in actual experiments when particles collide.

In units such that $c = 1$, the famous formula is just $E = m$. This robs it of its glamour, but says something simple and right. What you see in a particle, as its rest-mass, is the most you can possibly get out of it. There is no secret store of energy inside. As with other aspects of 'relativity' this is actually *more absolute* than pre-Einstein physics. To be more precise, Einstein did not actually write $E = mc^2$ in 1905, but he got within an ace of asserting this absolute and finite measure of potential energy.

Before Einstein's observation, exchanging energy was rather like going shopping: you could pop out and trade small bits and pieces, but no one had any idea what a house was worth. $E = mc^2$ tells you what the capital value of a particle is, and that it can buy a huge load of shopping. But it can only be tested by selling up and moving – actually changing from one rest-mass to another – and this needs the weak force which transmutes particles. In 1905 this was still poorly understood, but Einstein had a good guess: 'The mass of a body is a measure of its energy-content... it is not impossible that with bodies whose energy-content is variable to a high degree (e.g. with radium salts) the theory may be successfully put to the test.' For Forty years later that test was successfully made.

Nuclear bombs are not a direct consequence of realising the simple truth of $E = mc^2$. They also need the quantum mechanics of the nucleus. Arguably, $E = mc^2$ might not have been needed: the exponentially growing uranium-235 chain reaction could conceivably have been discovered without a full understanding of mass and energy. Certain uranium ores in Gabon formed a natural nuclear reactor two billion years ago; these might have shown the way. But in practice $E = mc^2$ did pave the way to nuclear fission. And Einstein himself started things off by writing to President Roosevelt in 1939 about the danger posed by German access to uranium. Thus began *the* superpower.

Time and tides

Popular science books also often say that Einstein's 1905 theory did not account for accelerations, but only dealt with constant motion. This is nonsense: his theory specifically concerned the accelerations induced by electromagnetic forces. What was missing in 1905 was a consistent account of *gravity*, and this is what Einstein supplied ten years later as the *general theory of relativity*. Again, the word 'relativity' is misleading. Far from 'everything being relative', Einstein's theory started out from the principle that physical reality cannot depend on how it is described, only on its absolute characteristics. It should be like classic Sudoku, pure pattern, independent of names.

Finding such absolute characteristics is not as easy as it sounds. The full theory of curved four-dimensional manifolds had taken decades for Gauss's successor Bernhard Riemann and others to work out. But it is not too difficult to see the general idea from the way that navigators use charts of the Earth's surface, making corrections for its curvature when calculating distances. Riemann's work showed how all such corrections could be brought systematically into a more general kind of Pythagoras's Theorem, still based on squaring. General relativity then incorporates and supersedes another famous square, Newton's inverse square law of gravity.

In this picture, gravity is not a force at all. Instead, the

Earth's mass curves the geometry of space-time in such a way that the Earth's surface is always accelerating upwards at 9.81 m/sec² and so presses on your feet. Weight doesn't exist, but the Earth's electromagnetic forces push harder on fat boys than on slim. This sounds crazy, but it is no crazier than the fact that if you steam straight ahead on a sphere you will end up back where you started. Such things are made possible by curvature.

Gravity, as space-time curvature, affects everything alike, and has a mutually attractive effect because there is no such thing as negative mass. That gives it a One-ness quite unlike electromagnetism with its cancelling plus and minus charges. A single particle only makes the most imperceptible change in the curvature of space-time, but gravitational effects are cumulative, and for a large body its One-ness outweighs and overwhelms the Two-ness of electromagnetism.

Yet the most important number involved in gravity is not One but Four. There are two ways to see why. One is to learn the quite tricky theory of curvature which was worked out by Gauss's successors. The other, a little easier, is to lie on the beach. Then you will notice the common-time rhythm of the tides every day: in, out, in, out. If you can see why the tides play a rhythm of four you are doing pretty well, because Galileo got the answer wrong.

If you made the very intelligent guess that it is four because there are four dimensions of space-time, you are, unfortunately, also wrong. The Four of the tides is a quite different Four. But the way that those fours coincide gives general relativity its very special character, in particular with the emergence of black holes where matter disappears from view and time comes to a dead stop. Einstein never believed in the reality of black holes; but his theory was more right than he was.

The structure of space-time, as elucidated by Einstein, could be called 'only a theory', but after nearly a hundred years it is as completely verified as anything in science. The effect of gravity in acting like a lens in intergalactic space is now as real as anything else in astronomy. A new space experiment is currently verifying the very subtle and small effect due to the Earth's rotation. Black holes are often described as mysterious objects found in space which baffle astronomers. This is the reverse of the truth: black holes were described in detail and extremely well understood before astronomical observations indicated that they actually exist. This understanding has relied on taking Einstein's theory far beyond its base in observations, but so far this extrapolation has been completely justified by new discoveries. Even so, physicists still find it hard to adjust to the universe as a full-scale curved space-time; it is all too natural to think of it as flat, needing merely minor corrections.

One reason why it is hard to adjust to four-dimensional space-time as a unity, as Minkowski wanted, is that it is hard to square with the apparent advance and asymmetry of time, so closely connected with the appearance of consciousness and freedom of will. There seems to be more to time than a minus sign. Minkowski wrote that time and space could be related by the square root of minus one, but although this points correctly to the emergence of complex numbers in modern theory, this doesn't help the reconciliation with conscious experience. It is even more difficult when time and space are muddled up inextricably in a *curved* manifold. Einstein's equations treat space-time as a big lumpy cake, in which time and space directions behave alike. Yet a conscious being seems to experience life like sliced bread, with one damned thing after another, and an intense asymmetry of future and past.

There is another missing link. Complex-number quantum wave functions do not, as they stand, fit into Einstein's account of gravity. Nor is it clear how Planck's quantum of existence, which applies alike to forces and particles, should extend also to the ripples of space-time. There are some incomplete pictures of how gravity and quantum mechanics combine. Stephen Hawking found a theory of how black holes can evaporate through quantum mechanics, and this suggests a connection between gravity, entropy and the quantum measurement operation. The ideal of a unified field theory, a theory of everything, as Einstein himself hoped for, still seems remote. But a deeper structure may emerge in some unexpectedly simple way. Hope springs four-dimensionally.

Foresight

Where might this deeper structure lie? Since the 1950s, ideas about time and space have moved even further towards thinking of *light rays* as primary. The rays at a point of Minkowski's space-time form a two-dimensional space, already introduced as the sphere of sight, or the sky. This is where the quaternions come back into the picture. The key idea is to combine them with Minkowski's observation about relating space and time by the square root of minus one. When this is done, they show how light ray directions can be thought of in a much simpler way. The sphere of sight can be identified with the complex numbers.

This should recall the electron with its spherical space of quantum states. This is the right connection, because the complex numbers perform an amazing double duty: they can define light ray directions and define electron states. This very unobvious structure, linking relativity and quantum mechanics in four dimensions, is rather like the hidden four-based structure of the integers.

These ideas use the equation 2 x 2 = 1 + 3 in reverse, finding a secret Two-ness in the one-plus-three of space-time. There is a much more radical step, which exploits this undercover connection to the full. It was discovered in the 1960s by the British mathematician Roger Penrose, at about the same time that he showed why the formation of black holes from collapsing stars was an inescapable aspect of general relativity. He gave his discovery the name of *twistor space*.

A complex number can label the sphere of light rays arriving at one point. To describe the space of *all* light rays, consider a light ray as a streak across space-time, and imagine that its track is actually visible. Choose any two points in space-time, and take a photograph of the sky at each. The track of the light ray will show up as a point on each of the skies. Each point determines a complex number, and the pair of complex numbers then gives an elegantly M-I-N-I-M-A-L description of the light ray. It is rather like the principle of surveying, or of binocular vision. The way that the brain receives a picture from each of two eyes, and puts them together, gives a rough picture of the Four-ness of twistor space.

There is a further step to take. To represent not a light ray, but a quantum photon, an extra 'twist' is required. It is achieved by a natural extension of twistor space to *four* complex numbers: this is a new Four, not connected in any simple way with the original four of time and space. The duality of electricity and magnetism, with its rhythm of four, is also naturally connected with these four dimensions.

There are two different ways of making this extension from light rays to quantum photons, a left-handed and a right-handed way. Choosing one rather than the other means breaking the symmetry of space and time. Twistors are there-fore asymmetric and so the weird asymmetry of the weak force may have a more natural expression in a twistor-based theory of physics. In fact, Penrose always hoped to use twistor space not just as another description of space-time, but to *replace* space-time as the fundamental description of reality.

Behind this two-blue-skies research there are strange, almost surreal ways of looking at spaces to discover hidden symmetries. Back in the 1960s a weekly British television comedy show, a sort of genteel precursor of *Monty Python's Flying Circus*, was called *It's a Square World*. At the time of the first manned space missions, it had a sketch where the astronauts looked out to see, to their surprise, a square world – or more precisely, a cube world. The kind of projection or perspective drawing that turns a sphere into a cube is a bit like the surreal transformations which bring to light unexpected properties of the physical world.

A simple example is that of *dilation* or change of *scale*. Putting a magnifying glass to a light wave – expanding both space and time by the same factor – gives a light wave of longer wavelength; it still looks like light. This is a size-doesn't-matter property which also applies to some aspects of gravity. It expresses something deep about the world which is more primitive than the metric – it does not need protons and electrons as clocks. One of Penrose's many ideas is that this hidden 'conformal' symmetry is fundamental, even though it is broken by gravity and by the restmass of particles. The symmetry is made transparent in the twistor description.

Absolutely fabulous four

The primary importance of light rays has been accepted generally in the development of fundamental physics since the 1950s. But few people took much notice of the extension of this idea to twistor space. In the opinion of vast majorities, the united theory of everything was expected to emerge from something quite different. This, also starting in the 1960s, was *string theory*. The fundamental idea is that if a quantum entity has a structure based on a *line*, rather than a point, then its track in space-time traces out a picture which is not like a tree, but a two-dimensional surface. This gives the possibility of uniting two (or more) kinds of Feynman accounts into one, very roughly:

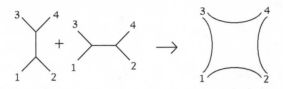

Strings gave a new way to put two and two together and make four. For some 30 years an enormous amount of work has gone into exploring these ideas, hoping to unite gravity and quantum field theory, and also to show the observed particles arising as harmonics played on this string instrument. But there was no apparent connection with Penrose's twistors. One reason for this is that twistors are absolutely based on the number Four, whilst string theory rejects the four-dimensionality of space-time, demanding that there must be further hidden dimensions.

But in 2003, the leading mathematical physicist Edward Witten drew together strings and twistors in a new geometrical picture, and applied them to something else: the colour force. Physicists had encountered considerable difficulty in analysing the three-based interactions of the colour force, but Witten showed how twistor geometry could usefully cut through the complexity as well as suggesting something more fundamental. For most physicists, unlikely to know anything about twistors, this was something very new.

It was also a revelation for me, because I had for some time been trying to express quantum field theory in twistor geometry, but had not realised how close the structures I had found were to the colour force problem. Essentially, the colour force can take the place of light, and instead of twistor space charting the possibilities for photons, it can do the same for the 'gluons' of quantum chromodynamics. One very important factor is that the *conformal symmetry* of the colour force is exploited in the twistor description, and this greatly simplifies the calculation of how it interacts.

On the edge

I began writing this book during a short visit to the Perimeter Institute, Waterloo, Ontario, Canada, where I was explaining these new pictures of the colour force. They result from taking the Feynman accounts, and translating them into twistor geometry using ideas which Penrose initiated long ago. They look remarkably like strings built out of squares.

This is a twistor picture of *four* interacting colour forces. It shows the trading of spin and colour rather than of energy, and makes a string-like surface rather than a network of lines:

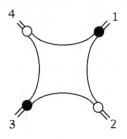

In more complicated interactions, millions of Feynman accounts boil down to sewing squares into a quilt, of which an example is:

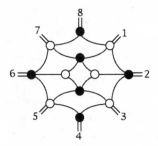

The expression coded by these pictures involves supersymmetry. In fact it needs *four* dimensions of supersymmetry. This further Four-ness is another link between twistors and strings.

These twistor accounts are not a complete theory – they are somewhere on the edge of providing a new kind of physical explanation. One important remaining problem is how to account for more advanced kinds of interaction where – to draw an analogy with the more advanced type of parties – uninvited guests turn up. These correspond to *loops* in the Feynman accounts, and *holes* in strings. But my belief is that these pictures offer a good clue to a deeper structure in quantum field theory.

The two times two of the twistor picture is incomplete, because it does not yet express the full content of Einstein's gravity. It is an inspired guess, showing how mathematical imagination can go beyond the immediate Four-ness of the world to –

5

Fifth Freedom

'Freedom,' wrote Winston Smith, 'is the freedom to say that two and two make four. If that is granted, all else follows.' But under torture, he says: 'Five! Five! Five!' George Orwell, like G. H. Hardy, held on to mathematics as a paragon of objective truth. That two and two make four was his version of 317 being prime, because it is so, whatever anyone tells you to say.

Like Hardy, Orwell was sceptical about the practical role of science, Hardy's gloomy verdict on radio being that it had resulted in menacing propaganda, Orwell foreseeing that the main aims of science would be 'to discover, against his will, what another human being is thinking, and... how to kill several hundred million people in a few seconds...' Orwell's language of Newspeak might well have been influenced by the functional and nuance-eliminating artificial language Interglossa, devised and published by none other than Lancelot Hogben in 1943.

I wonder what Hardy and Orwell would have made of Big Brother showing that in reality, people will do and say anything merely to attain celebrity – all amidst rolling news info-tainment so readily absorbing the torture of unpersons, with language like 'rendition' surpassing Orwell's cynical ingenuity. Five is the new four, no problem! Actually *Nineteen Eighty-Four* does show the author's grumpy old man aspect, despairing of those who would not take the world seriously enough. Perhaps Orwell also took an unrealistically high-minded attitude – Hardy's attitude – towards objectivity, for fiction and faith are generally valued more highly.

Science is less popular than reality shows, and magic has more mileage. Five is the traditional number of magic, which has just the same inquisitive roots as science, but is more comfortable with putting two and two together and making – well, 575. Unfortunately the very things people want most from science may prove the hardest to tackle. The success of classical astronomy, with its wonderful predictions of eclipses, gave a misleading picture because they involve spaces of so few dimensions. Even Newton, having explained the motions of the solar system, got nowhere with the complexities of chemistry. Scientific medicine is very recent, and with epidemics and earthquakes science still disappoints with its inability to give exact predictions. Climate prediction has only just arrived inside the borderline, thanks to the possibility of huge computer-based trials.

What's more, a magically inspired guess, based on very little data, is sometimes right; this was true with Fermat's conjecture, with general relativity, and the colour force. Popper's famous criterion of falsifiability is not the story of how science actually arrives at new ideas: personal, cultural and aesthetic criteria come into it. String theory and twistor theory are remote from experimental test even though ultimately that is how they will live or die. Mathematics and mathematical physics (often to the annoyance of experimental scientists) are especially influenced by aesthetic criteria of elegance, and there may be little agreement on when those criteria are met. The actual business of research is a very subtle relationship between fact and imagination.

With this in mind, I take the number Five more positively than Orwell did: not as an image of falsity, but as the number of imagination, guided by aesthetics as well as by contact with reality. This reflects the ancient idea of four earthly elements and a fifth, 'quintessence', for the celestial regions. G. H. Hardy in 1940, at a time of crisis when everyone was supposed to be practical and pulling together, asserted a fifth freedom: not just the freedom to say that two and two make four, but to roam in pure mathematics, not knowing where it might lead. President Roosevelt proclaimed four freedoms a little later, and the 'fifth freedom' has been evoked by others to express more challenging aspirations. Asserting mind over matter, as do disabled people (including the President himself, though secretly), is perhaps the most genuine magic of human life. Stephen Hawking has shown the power of the inner eye, exploring space-time despite intense physical limitations. It is the eye of mathematics, seeing with insight.

Magic squares

Some squares are actually called magic, and were known in ancient Chinese, Egyptian and Indian mysticism. The magic consists of having the rows and columns all adding to the same total, and the diagonals as well. The 3 x 3 magic square, using the numbers One to Nine, is:

$$
\begin{matrix}
6 & 1 & 8 \\
7 & 5 & 3 \\
2 & 9 & 4
\end{matrix}
$$

You can easily show by Sudoku methods that there is essentially only one such square – that is, you can obtain all the others by permuting the rows and columns. There is no 2 x 2 magic square. There are 880 different 4 x 4 squares and 275305224 5 x 5 squares. The general problem of counting them is not solved.

In 1776, Euler showed a way of generating magic squares from two Latin squares. In the 3 x 3 case, this amounts to writing down

1	0	2
2	1	0
0	2	1

2	0	1
0	1	2
1	2	0

then combining them into

12	00	21
20	11	02
01	22	10

and interpreting the entries as numbers written in base 3. Add 1 to each cell to make the numbers run from 1 to 9 rather than 0 to 8. This results in the magic square above: the addition of all rows and columns to 15 comes out automatically through this construction.

Euler's idea does not solve the problem as there are many magic squares which do not arise from two Latin squares in this way. But it illustrates a very general mathematical strategy of finding a way to break a structure down into a multiplication of simpler things. The splitting of on-beat primes into complex pairs, the understanding of the proton as three quarks, and the description of light rays with complex numbers, are successful examples of this kind of generalised factorisation.

Magic squares will probably not lead to any more serious discovery. Yet Euler's observation is a model of other explorations which at the time might have seemed self-indulgent play with useless abstractions. The quaternions discovered by Hamilton, like his development of Newton's laws of motion, only came to life many decades later when they became vital in quantum mechanics. The geometry of many-dimensional curved spaces came out of Riemann's work in the 1850s, which was not driven by any practical need, but now through general relativity keeps satellites in place. In this spirit we step ahead from four to five by starting on the magic pentagon.

The line of beauty

A pentagon has a natural way of growing new pentagons. Drawing the interior diagonals and exterior extensions turns it into a star, as found in ancient magic formulas:

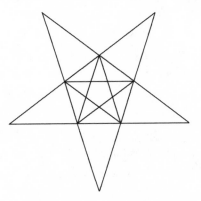

The five points of the star form a new large pentagon, of which the star gives the diagonals. A new smaller pentagon emerges at the centre, for which new diagonals could be drawn, and so the whole process be repeated indefinitely.

A natural question to ask is how large the new pentagons are, compared with the original one. We can also ask about the lengths of the diagonals. Both questions can be answered together by starting with a simple observation. There are only two triangular shapes in the picture: fit and fat. That is, they have the same shape, although they come in different sizes.

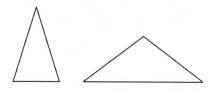

Inside every fat one there's a fit one, struggling to get out, leaving a smaller fat one.

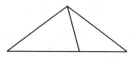

Less well known is that inside every fit one there's a fat one, leaving a smaller fit one.

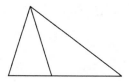

Take the sides of the original pentagon to be 1 unit. We shall use the Greek letter ϕ (phi) for the lengths of its diagonals Extract these two triangles from the star:

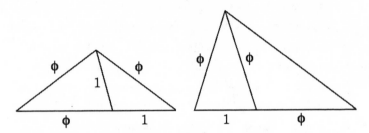

The first figure has the fat shape, with lengths greater by a factor of ϕ than those of the fat triangle which is inside it. Thus its base must be of length ϕ^2. Yet its base is also $1 + \phi$. So ϕ must have the special property that

$$1 + \phi = \phi^2.$$

The second figure, with the fit shape, tells the same story, which is also equivalent to

$$\phi - 1 = \phi^{-1}.$$

The ratio $1 : \phi$ is the same as the ratio $\phi - 1 : 1$, and it is the ratio of every two consecutive line segments in the five-star picture. You could now look for ϕ using a calculator: it is a number which differs from its reciprocal (and from its square) by exactly 1. If you have ever noticed that the conversion factors for miles to kilometres and vice versa are close to this (being 1.609 and 0.621 respectively), then you have a start. By experimenting with a calculator, you can find the approximate values:

$$\phi = 1.61803398875, \phi^{-1} = 0.61803398875,$$
$$\text{and } \phi^2 = 2.61803398875.$$

We can do better than this by finding a connection of ϕ with 5 – not entirely surprising, since it comes from a pentagon:

$$(\phi + \phi^{-1})^2 = (2\phi - 1)^2 = 4\phi^2 - 4\phi + 1 = 4(1 + \phi) - 4\phi + 1 = 5,$$

and so

$$\phi = \tfrac{1}{2}(\sqrt{5} + 1).$$

This deconstructs the secret: the magic comes from the irrationality of square roots. But there is more to it.

Cool as phive

The ratio ϕ is called the *golden* ratio, mean, section or number. A rectangle with this ratio for its sides has the property that if you remove a square from one end, the remaining rectangle is also golden. By continuing the process, it gives a spiral of squares:

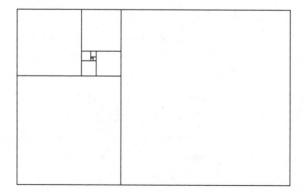

A rectangle of this shape has been held to be the most pleasing: not obtrusively long, not boringly squat. Plato thought it was just perfect. I am a sceptic about such a classical ideal, not because I consider A4, portrait, landscape or widescreen rectangles to be better shapes, but because such a fixed aesthetic seems incompatible with the ever-changing development of human vision and expression. The romantic ideal was quite different: Hogarth taught an S-shaped line of beauty. Current conceptual art abandons beauty and all authority, the only rule apparently being to do something new, however trivial.

There is a three-dimensional analogue of the golden number. Whilst the golden ratio is the number which is squared by adding one to it, the *plastic number P* is cubed by adding one to it: i.e. $P^3 = P + 1$. Instead of being based on 5, it is based on 23, the exact value being

$$P = (1/2 + 1/6\ \sqrt{23/3})^{1/3} + (1/2 - 1/6\ \sqrt{23/3})^{1/3}.$$

Numerically the plastic number is 1.324718..., nearly 4/3. (Approximating it by 4/3 is the same approximation as is made in music by taking the seventh harmonic to be equal to two fourths.) The 23 arises as $3^3 - 2^2$, parallel to $5 = 2^2 + 1$. Does this give the number 23 some aesthetic property? David Beckham, who took this number, is beautiful, but the mystical cult of 23 finds sinister characteristics in it.

The exact formula for P, with its cube roots, tells another important story about Five. It is obtained by solving the cubic equation exactly, not an easy business, and the greatest advance in European mathematics before the scientific revolution. The quartic equation $x^4 = x + 1$ can also be solved in a similar way, but the quintic equation $x^5 = x + 1$ cannot be solved by a method which needs only adding, subtracting, multiplying, dividing and the taking of roots. The non-existence of such a classical method was established by Abel and Galois, both very young, in the 1820s. As so often, proving the impossibility of something led to completely new advances. Quintic equations, ugly customers to the eighteenth century, became a gateway of enlightenment.

Aesthetic criteria change: the fifth harmonic gives the major third, a paragon of concord in the common chord, but now for art-music almost unbearably – well, common. Beethoven's Fifth Symphony resolves into a triumphant major C-E-G, but when Vaughan Williams opens the last movement of his 1935 Fourth Symphony with three major chords, it speaks of horrible irony. Many suspect the same of Shostakovich's Fifth. The blue notes of jazz likewise do everything to avoid asserting that fifth harmonic. Only genius can still make a major third sound new and cool: Messiaen, Pärt, even Glass now and then, and of course the Beatles. Nothing can stop the ear from hearing the resonance of the fifth harmonic, but it does not supply a static standard of perfection, only a stimulus to ever-new forms of expression.

Breaking down the fat

We return to the shapes with the Plato-rated ratio, and use them to extract a sequence of numbers. First, we arrange the shapes into a sequence, beginning as follows and extending indefinitely from small to medium, large, XL, XXL...

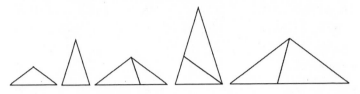

Each fit and fat pair combine to make the next triangle in the sequence, or equivalently, each triangle breaks apart into the preceding fit and fat. Note that there are two ways of making the breakdown.

Each triangle has an area exactly ϕ times that of the preceding one, so we have another example of exponential growth. The A paper series simply gives the power of 2 arising from one shape, but something more subtle emerges from counting these two different shapes.

Suppose we break all the triangles down into the smallest-size fit and fat triangles. This can be done in many ways, but if we are only counting the number of shapes this makes no difference to the answer. So, for instance,

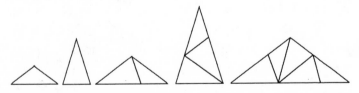

shows that the fit/fat count is, respectively: (0, 1), (1, 0), (1, 1), (2, 1), (3, 2). Each one comes from putting together the previous two shapes, so the sequence will continue (5, 3), (8, 5), (13, 8), (21, 13), (34, 21)...

Rabbit family values

From now on we will concentrate on the properties of these integers, which are very famous in mathematics. They define the *Fibonacci number* sequence

$$0, 1, 1, 2, 3, 5, 8, 13, 21, 34, 55, 89, 144\ldots$$

in which each number is the sum of the previous two. It was the same Fibonacci who appeared in Chapter 1, introducing the base-10 Eurabian numerals. This sequence of numbers had been known and studied in classical Indian rhythms, but Fibonacci described them in a more down-to-earth way, in terms of rabbit propagation. I will refrain from drawing a family tree of the inbreeding bunnies, leaving these activities to the reader's imagination. My fit-and-fat pictures are intended to be more tasteful than Fibonacci's cunnilinguistics, although of course lines of beauty are in the eye of the beholder, and one man's meat is another man's bottom line. The noughth Fibonacci number is 0, and the next 25 are:

1	1	6	8	11	89	16	987	21	10946
2	1	7	13	12	144	17	1597	22	17711
3	2	8	21	13	233	18	2584	23	28657
4	3	9	34	14	377	19	4181	24	46368
5	5	10	55	15	610	20	6765	25	75025

At first sight they look completely random, but a second look shows a hidden connection with the primes.

Five-finger exercise

The number Five makes this connection most transparent. The fifth Fibonacci number is 5. Then every fifth Fibonacci number thereafter is divisible by 5. Why is this, and how can one be sure that the pattern will continue forever?

One way is just to look at the *last figure* in the Fibonacci numbers. These run

(1, 1, 2, 3, 5, 8, 3, 1, 4, 5, 9, 4, 3, 7, 0, 7, 7, 4, 1, 5, 6, 1, 7, 8, 5, 3, 8, 1, 9, 0, 9, 9, 8, 7, 5, 2, 7, 9, 6, 5, 1, 6, 7, 3, 0, 3, 3, 6, 9, 5, 4, 9, 3, 2, 5, 7, 2, 9, 1, 0)

and that sequence of 60 will then repeat. Every fifth number is either 5 or 0. However, it is more systematic to use the language of congruence and modular arithmetic introduced in Chapter 4.

First, look modulo 2: i.e. at evenness or oddness. The sequence runs: odd, odd, even, odd, odd, even... which we can write as (1, 1, 0, 1, 1, 0, 1, 1, 0...) in a repeating cycle of three. Modulo 3, the repeating cycle turns out to be of length 8: (1, 1, 2, 0, 2, 2, 1, 0). Modulo 4, it is of length 6: (1, 1, 2, 3, 1, 0).

Modulo 5, the repeating sequence is (1, 1, 2, 3, 0, 3, 3, 1, 4, 0, 4, 4, 3, 2, 0, 2, 2, 4, 1, 0) of length 20, and this is the simplest way to see that every fifth Fibonacci number must be divisible by 5.

The same applies to any modulus. Extending this argument a little, you can see that because the seventh Fibonacci number is 13, every seventh number is divisible by 13. In general, the nth Fibonacci number F_n must divide exactly into all of F_{2n}, F_{3n}... If m and n have a common factor p, then F_m and F_n share F_p as a common factor.

Break the real Fibonacci code

These more challenging problems can be solved by extending these counting arguments. They reveal more of the way that the prime numbers are hidden inside the Fibonacci numbers.

TRICKY: Show that every prime is the factor of some Fibonacci number. Suppose a prime p first occurs as a factor of F_s, then $s \le p^2 - 1$. (Hint: consider how many pairs of numbers there are which could be consecutive integers in the Fibonacci sequence modulo p.)

FIENDISH: In fact $s \le p + 1$.

SUPER-FIENDISH: In fact s is a factor of $p + 1$ if p ends in 3 or 7; s is a factor of $p - 1$ if p ends in 1 or 9. (Hint: the question turns on whether there exists a discrete ϕ modulo p, i.e. a number f such that f^2 is congruent to $f + 1$ modulo p. This can be solved by using Gauss's Law of Quadratic Reciprocity.) Five is special, as F_5 is 5.

Neighbours from hell

Neighbouring Fibonacci numbers, or those two apart, have no common factor. Those three apart can at most share a factor of two. You might compare the walk the sequence takes from 3 to 55 with the route that the triangular numbers take – 3, 6, 10, 15, 21, 28, 36, 45, 55 – which has common factors as intertwined as the residents of Wisteria Lane. The Fibonacci numbers are like spoddy webcammers, chatting globally, but unaware of their local neighbours. (There is a sense in which the Fibonacci neighbours are as lacking in common factors as they could possibly be, which will be explained in Chapter 9.) Far from exemplifying classical harmony, they are more like a lumpy installation at the Tate Modern.

The Fibonacci sequence has unique properties which makes it pop up throughout mathematics. But it does not stand alone. There is one obvious thing setting the sequence in a larger context: it grows exponentially. The ratio between neighbouring numbers gets closer and closer to ϕ, as you can see from their origin in the sequence of triangles where the growth ratio is exactly ϕ. It is not true to say that this sequence is the only way to arrive at ϕ. You can start with any two numbers you like and use the same addition rule for generating a sequence: the golden ratio will appear as the limiting growth factor. However, the Fibonacci sequence has an extra symmetry, and a graph of the odd Fibonacci numbers lies on exactly the same exponentially growing shape as the soap film of Chapter 2. This symmetry and the exponential dependence on ϕ are shown by the formula

$$F_n = \frac{1}{\sqrt{5}}\left(\phi^n - (-\phi)^{-n}\right) = \frac{1}{\sqrt{5}}\frac{(1+\sqrt{5})^n - (1-\sqrt{5})^n}{2^n}$$

which is useful, but does not make it obvious that each F_n is an integer! This is typical in the aesthetics of mathematics: a formula or picture illuminates one aspect of a structure, yet disguises another one. The mind needs many different pictures to build up understanding, piecing them together into a manifold of insight.

As integers, the F_n have Diophantine relationships similar to those of the numbers used to calculate $\sqrt{2}$ in Chapter 4. The square of F_n differs by 1 from the product of its neighbours. F_{2n-1} is the sum of the squares of F_n and F_{n-1}, and $F_{2n} = F_n \times (F_{n-1} + F_{n+1})$; these are analogous to the jumping-ahead rule for approximations to $\sqrt{2}$. In fact the fractions $(F_{n-1} + F_{n+1})/F_n$ give better and better approximations to $\sqrt{5}$, and satisfy

$$(F_{n-1} + F_{n+1})^2 = 5 \times F_n^2 \pm 4.$$

There is a duller dual sequence $(0, 1, -1, 0, 1, -1, 0...)$ in which each number is the *negative* of the sum of the preceding two. You would be right to suspect that its repetitive cycle is an aspect of the duality of growth and oscillation. In fact any recursive rule of this type can be reduced to such growth and oscillation: this is an aspect of the fundamental theorem of algebra for complex numbers. The plastic number comes into another very special example of this. It is the limiting growth ratio of the sequence

0, 2, 3, 2, 5, 5, 7, 10, 12, 17, 22, 29, 39, 51, 68, 90, 119, 158, 209, 277, 367, 486, 644...

You are invited to find the rule that is being applied to generate this sequence, and to spot a magical connection with the prime numbers.

Artists, architects, musicians, not to mention fiction-writers, like to bring Fibonacci numbers into their work, and such borrowing is generally admired. But I doubt whether the numbers *in themselves* lend any aesthetic character to artistic work. Copying them is no more worthwhile than chanting a multiplication table. The value and beauty, if any, lies in the *understanding* of the sequence, with its roots in five-fold symmetry, the interlocking with primes, the connection with growth and rotation, and the infinitude of all these and many other properties.

Crazy paving and crystal maths

The broken-down triangles suggest that there is a further story to be found. Is there some systematic way of doing this breakdown in a way that respects the self-similarity and five-fold symmetry of the original pentagram? The *Penrose tilings of the plane* achieve something very closely related to this. These were discovered in the 1970s by Roger Penrose, the originator of twistor geometry.

In the best-known form of the construction, the fat and fit triangles are doubled up into *kites* and *darts,* thus:

The arrows indicate a constraint: that two tiles can adjoin only if the arrows agree. A very rich body of theory then shows that with this constraint the entire infinite plane can be completely covered with the kites and darts in such a way that the tiling is *aperiodic*. Wallpaper is periodic: there is always a way of moving across the wall and up the wall by certain fixed lengths, so as to arrive at another point where the pattern is exactly repeated. There is no such symmetry in a Penrose tiling. The reason for this rests on self-similarity and the golden ratio, the concepts found in the pentagram.

A very small piece of tiling looks like this:

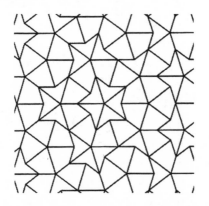

You can check that the arrows are consistent with these shapes. This fragment also illustrates a very important property: the freedom in creating a pattern. Starting from the central star of five darts, a choice *could* have been made to keep the five-fold symmetry going. Instead, by choosing two kites rather than two darts, the symmetry

165

is broken. In covering the whole plane, infinitely many choices must be made, and there are indeed infinitely many different coverings of the plane. In placing the tiles consecutively, however, it is very far from obvious where there is a freedom to make a choice and where there is not. This problem of choice and rules is in fact the main point of the Penrose tiling.

In early 2007, a paper in *Science* by the physicists Peter Lu and Paul Steinhardt analysed medieval Islamic mosaics where pentagonal symmetry plays a stunning role in the design. Of one particular work, the 1453 Darb-i Imam shrine at Isfahan, Iran, they wrote that the designs 'were nearly perfect' Penrose tilings, although 'the underlying mathematics were not understood for another five centuries in the West'. This claim was taken up enthusiastically in the media. But according to the analysis itself, the designers were placing tiles without respecting the 'matching rule' of the arrows. And this rule is not just an 'approach' to creating patterns as these writers suggest: it is the *definition* of the pattern, just as having every number from one to nine in every row, column and subsquare is the definition of Sudoku. Although it is conceivable, as they suggest, that workmen made errors in placing the tiles, it seems to me virtually impossible that the designers had the modern theory.

The mathematics behind the Penrose tiling, which took another five centuries, is not really about pentagons. Fiveness is the medium for a deeper mathematical message about infinity and about unity, a message which comes from modern work starting from Gödel. Penrose tilings give a way to embody the logic of aperiodicity which is, as so far known, the simplest possible. But the logical message can be conveyed equally well by shapes which have nothing to do with pentagons.

This does not make the medieval tilings any less marvellous. As images of inspiration, as well as meticulous constructions, they come as a shock to modern European eyes, and a reminder that al-Khwarizmi's mathematical culture was so far ahead.

Another shock, and a parallel question about rules, arises in asking whether Penrose tilings can appear in physical reality. So far we have had various deep levels in which integers play a role: symmetries of space, the quarks, the nuclear numbers, the electrons and their bondings. There is another level on a scale higher than that of atoms: crystals. For very good reasons, five-based structures had never been expected in crystallography. Four or six, yes: five no. Yet in 1984 – as it happened – solid-state physicists were in for a big surprise. Pentagonal patterns had been discovered in certain metal alloys. To some it was like hearing that two and two made five, and the news was much resisted.

These aperiodic solids are called 'quasi-crystals' and do seem to be three-dimensional forms of Penrose tilings. As always, quantum mechanics is involved. Penrose's question about the rules for constructing the tilings leads to the question of how the atoms in a growing quasi-crystal can possibly manage to situate themselves so as to create a globally consistent pattern. Much more is now known about electronic structure in quasi-crystals – but there does not yet seem to be a clear answer to this question, which relates to the most fundamental questions about quantum mechanics.

Kurt Vonnegut's story *Cat's Cradle* was a warning about the scientific imagination. Vonnegut challenges the Hardy-like claim that the beautiful and true demands to be followed regardless of consequences. Quasi-crystals, as surprising in reality as the 'ice-nine' of Vonnegut's fiction, make a test case So far, they have only been used for creating slippery surfaces, and I am unaware of any danger posed by their discovery. But if there were some malign use, it is hard to see how it could have been anticipated and averted. Where and when was the place to draw a line on a slippery slope? Was it in Plato's aesthetics, Euclid's pentagons, Fibonacci's rabbits, or in the inspiration of Islamic art and its parallels in Kepler? Was it in Gödel or in Wang, the pure logician who started the link from logic to tilings? Did it happen when Penrose refined the aperiodic tilings to kites and darts, or when another mathematician, Robert Amman, found similar patterns independently? Nor is there a simple way to draw a line between mathematical theory and physical discovery, because interpretation of the X-ray pictures needs considerable mathematical understanding.

There is a yet higher level where the integers play a role: life.

On the ground

The Fibonacci numbers appear quite conspicuously as integers in the leaf and flower patterns of certain plants. They have been known for centuries, and for this reason, have turned up relentlessly like C-list chat-show guests in books about numbers and science, not making a very good advertisement for mathematics because they have only been observed and not fully understood. Life is the hardest area for mathematics to exercise explanatory power, and there is much danger of jumping to wrong conclusions. Many more people now know of Fibonacci numbers, because of the Da Vinci Code fiction. But this has exaggerated their significance in a magical direction: Fibonacci numbers are not the key to all Nature; they just occur in certain particular physiological processes.

Sunflowers make for the clearest examples of Fibonacci numbers in the spiral patterns of their gigantic seed-heads. They can have up to 89, 144, 233 in their patterns. In Tony Robinson's most recentl archaeological series he debunked *The Da Vinci Code*. He had a scientist pooh-poohing everything about the golden ratio in biology. But Fibonacci leaf and bud arrangements have been meticulously measured and documented. This programme missed an opportunity to explain a *genuine* puzzle of numbers of the kind that hooks people on science, as opposed to the Dan Brown principle of adding two and two to make a profitable five. In 1995 some first good explanations were given of why Fibonacci numbers would occur. Their stand-offish, repulsive, un-neighbourly characteristics are the key.

There is an evolutionary factor favouring Fibonacci numbers in plants. Their unneighbourly property is highly suitable for leaves which must sprout with equal weight and light catchment in all directions. But the number 233 as against 234 in giant sunflowers is far too specific to be from evolutionary selection alone: it must be the outcome of an actual physical aspect of the growth process. Evolution can only exploit the pathways made available by physical reality. It seems that Fibonacci development is, in some processes, a physical constraint. But that actual physiological mechanism is still lacking. Generally speaking, there is no complete explanation of the whole process from coded genes to proteins via chemical reactions to the growing shapes of organisms.

There are broader questions prompted by the inelegant, gangly sunflowers, which are also famous because of Vincent van Gogh's almost painfully unpretty paintings. Did Vincent notice the Fibonacci patterns? It appears not: he seems to have ignored them in a broad-brush treatment. Detailed representation of physical appearance was not his purpose. As photography made three-into-two art redundant, *fin de siècle* Europe was bursting with new vision. The very idea of symbolism and representation took on new life, and so it did in mathematics. It was during this same period that mathematics finally separated abstract number from physical space. It was also at this time that physics reached beyond the immediately visible and measurable to the abstractions in relativity, radioactivity and the quantum. Proust loved to be compared with Einstein, both upsetting received views, not clinging to ancient standards of perfection. Van Gogh's harsh goldenness anticipated the coming twentieth century. The five-four waltz in Tchaikovsky's last symphony signals an end to classical rhythm, and Holst's 'Mars, the bringer of war', so prophetic in its ugliness for 1914, is also in five-four time.

The sunflower tells a story about quantum mechanics. It is striking that green leaves and yellow flowers, by virtue of their colour, are rejecting light from the centre of the Sun's spectrum – apparently as ungrateful as young humans spurning nutritious food with disgust and insisting on a diet of fat, sugar and salt. A matt-black plant, like a solar panel, could absorb much more energy. Sunflowers, turning towards the Sun yet throwing away its most brilliant light, seem to exhibit evolutionary fatness, not fitness! But green is cool: if plants were black they could not use the energy effectively. For this they depend on the quantum magic of electrons in the chlorophyll molecule – as indeed the molecules of the retina by which we see 'green' also depend on quantum mechanics.

Cultiver nos jardins

Biologists tend to be suspicious of grand mathematical arguments and physical principles, given the incredible complexity of detail connecting the genes, chemistry and environment of an organism. But possibly there are mathematical constraints on physical form which cut through the complexity and favour – for instance – the five-fingeredness which goes far back in evolution. If so, your present location just halfway through *One to Nine* is not just a consequence of an arbitary convention to write numbers in a scale of ten digits, but goes much deeper into the very nature of physical embodiment. Alan Turing, already quoted as talking about conjectures, started a ground-breaking investigation of such ideas in 1950. (It was literally ground-breaking as he ran round collecting plants, and Turing was literally running round, as he was a champion marathon athlete.) Few knew of his strong interest in Fibonacci numbers in plants, because he left this work unpublished at his death in 1954. What he did publish was a much broader theory of how biochemistry could give rise to the shapes and patterns of biology, based on the mathematical properties of equations which became accessible to study through the then new electronic computer – essentially his own invention.

Since then, Roger Penrose has contributed a quite different argument that there is a physical grounding in quantum mechanics to the brain, on which consciousness depends. He has suggested a crucial role for structures called *microtubules*. One striking feature of the microtubules is that they have a pattern as definite as that of the sunflower heads, based on the Fibonacci numbers 5, 8 and 13. If Penrose is right, Five is at the root of consciousness.

This is not at all the prevailing view of scientists in neurology, neural networks and Artificial Intelligence, and has gained no general acceptance. But it is still unclear what the physical basis of memory actually is. When learning that two and two make four, not five, what is actually happening in the brain? Coming back to *Nineteen Eighty-Four*, what is it that makes seeing the *truth* of $2 + 2 = 4$ so different from rote-learning $2 + 2 = 4$, a sequence of syllables that might just as well have been $2 + 2 = 5$? This is Penrose's central question. In his book *Shadows of the Mind* he has suggested a new view of the connection between the mental and the physical, much more subtle than the false choice between Hardy-ish idealism and Hogben-ish practicality.

There has recently been a dramatic development. Late in 2006, a group of European molecular biologists reported a first example of where Turing's theory could be verified with actual chemicals reacting in just the way he had suggested. It was found in patterns for mouse hairs. So there is at least one place where his conjecture seems right – and there is no reason to suppose it a unique case. Some 54 years have passed since Turing saw this principle of morphogenesis and worked out the theoretical framework. There is plenty of scope for more discovery by the mathematical eye. Meanwhile there is a lot of life in the number –

6

The Joy of Six

Amo, amas, amat; amamus, amatis, amant. Three persons, each both singular and plural, make Six. For Latin lovers, Six is sex, and in *soixante-neuf*, even the numeral is erotic. Those Indo-European roots suggest that Six, like sex, is connected with cutting and dividing. Six is the first number with more than one factor, capable of division in different ways. It is the number of choice, change and chance.

Life depends on the nucleus with six protons, surrounded by an outer electronic shell which is just half-empty, half-full. The twos and fours of quantum mechanics allow the carbon atom to bond to oxygen. But carbon dating works energetically with other atoms, with just the right chemistry for the complex molecules of life. Equally penetrating and penetrated, carbon atoms bind happily to other carbon atoms, with what in the Sixties was called polymorphous perversity.

Diamonds may be your best friend, but there are alternative forms of bonding. In 1985 a new form of pure carbon, called C_{60} for its 60 atoms, was identified. It's odd that it took chemists so long to spot this molecule, because, like quasi-crystals, it had already been anticipated in theory, and because the polyhedral geometry was so familiar after the 1970 World Cup. Many had been getting their kicks with Six:

Every carbon atom in a molecule of C_{60} lies on one pentagon and two hexagons. It turned out that this secret passage from Five-ness to Six-ness had been sitting in burnt toast all the time, unrecognised.

A twisted mind

Six is two threes, and six is three twos. The multiplication of numbers is insensitive to order: first doubling and then tripling is the same as first tripling and then doubling. If we add further structure, two threes are not the same as three twos. In music, Holst evokes mercuriality in 'Mercury' by suddenly shifting from 2 × 3 to 3 × 2 rhythm (and also hopping unexpectedly by six semitones from B♭ to E). This is rather contrived, but more natural three-against-two effects are common in Latin American dance. As a crossover compromise, *West Side Story* has the cross-rhythm of 'I want to be in America'.

There is an analogue in a sort of twisted multiplication which first occurs when Six comes into the picture. There is a problem in geometry that could crop up when assembling furniture from a flatpack. Suppose you have a flat triangular piece which has to slot into the rest of the design, but you cannot see which way round or even which way up it goes. There are six possibilities to try out. On top of the general difficulty of following flatpack instructions, an inescapable mathematical fact can make this totally confusing. Rotating the triangle, then flipping it over, does one thing. But giving it the flip first, and then the rotation, ends up with something else. Operations, unlike numbers, are sensitive to order. As soon as Galois, Abel and Hamilton started writing symbols for operations, a serious new algebra began with twisted multiplication: $a \times b$ may not be the same as $b \times a$.

We have already touched on this twisting with the quaternions in Chapter 4. They have just this 'non-commuting' property, because they represent rotations in three dimensions. The flatpack operations can be thought of as picking just six quaternions which form a closed *group*. The question of how many such sets of operations exist, starts off a huge area of modern algebra which must be indicated with just a few images. One picture is of the *crystals* and their symmetries in three dimensions. Crystals are classified by rotation and reflection symmetries, and the reason why quasi-crystals were such a surprise, is that being aperiodic they failed to fit into this classification. But crystals are only the start: there are rotations in spaces of any number of dimensions, and the analogues of crystals. The non-commuting of rotations in the quark's colour space is an essential feature underlying the confinement of the nuclear force. This in turn points the way to the geometry of string theory, where physicists hope to find symmetries related to the fundamental particles and forces.

There is a wonderful image of the complexity of such questions in the *Rubik cube*, with its non-commuting twisting operations. It was the Sudoku of the 1980s. For the standard 3 x 3 x 3 Rubik cube there are, it turns out, 43252003274489856000 possible positions. Solving the puzzle amounts to finding a path from one position to another amongst this number, the order of twisting operations being crucial. Like Sudoku, it also gave a thought-provoking image of how fascinating logical puzzles can be, in a way that official education completely misses. Schools banned it, and young people developing an amazing intuition for its geometry were laughed at.

Meanwhile, in the 1980s, algebraists were finally sorting out what turned out to be a kind of Rubik cube at the centre of mathematical reality. There are certain structures, enormous extensions of the flatpack problem, which are to rotations rather as the primes are to numbers: these are the *finite simple groups*. They are very special, definite, extremely large and complicated structures, still full of mysteries and new discoveries. They culminate in the Monster group which has 2^{46} x 3^{20} x 5^9 x 7^6 x11^2 x13^3 x 17 x 19 x 23 x 29 x 31 x 41 x 47 x 59 x 71 = 80801742479451 2875886459904961710757005754368000000000 elements, rather than the six of the flatpack.

Sudoku is a very much simpler problem, but it still presents huge numbers of ways in which the numbers 1 to 9 can be combined, to be sorted out systematically. The number Six is a good starting point for seeing how changes and choices combine. They naturally lead on to the questions about chances. Less obviously, they will also lead back to the duality of growth and oscillation, and from there to primes, powers, dimensions and spheres, and then back to life.

Ringing the changes

In Killer Sudoku you may find a box of three cells where the numbers must add to 6. The numbers then must be 1 + 2 + 3 = 6 (thus showing 6 as a triangular number). But the 1, 2 and 3 may be in any of the six orders: 123, 132, 213, 231, 312, 321.

There are six orders because 6 = 1 x 2 x 3, and this fact is what makes 6 the starting point for permutations and combinations. Before embarking on this, it's worth remarking on the coincidence that

$$6 = 1 + 2 + 3 = 1 \times 2 \times 3$$

This property is tantamount to the fact that six is what the Greeks called a *perfect number*. Constance Reid chose this feature for Chapter 6 of *From Zero to Infinity*, leading into an unexpected tale about the 'Mersenne primes'. We will return to this in Chapter 8. In this chapter, we explore the complexity rather than the perfection of Six.

We say that 6 is '3 factorial' and write 3! as shorthand for 1 x 2 x 3. Why is this the number of orderings of three things? It is easy to write them out and count them, but it is better to argue systematically. The first can be 1, 2 or 3; once this is chosen there are two possibilities for the second, and then there is only one choice for the third. This is like a tree with three branches, each dividing into two, and each of these ending in one leaf: this gives a total of 3 x 2 x 1 leaves. The same argument works for the number of ways of ordering four things:

$$4! = 4 \times 3 \times 2 \times 1 = 24$$

Instead of working these out from scratch, we can recycle the work already done: 4! is 4 x 3!, and then 5! is 5 x 4!, and so on:

$$5! = 5 \times 4 \times 3 \times 2 \times 1 = 5 \times 4! = 120$$
$$6! = 6 \times 5 \times 4 \times 3 \times 2 \times 1 = 6 \times 5! = 720$$
$$7! = 5040, \; 8! = 40320, \; 9! = 362880.$$

This last number gives the total number of possible rows (or columns, or subsquares) of Sudoku.

You can get a picture of these numbers from the task of ringing the changes on a set of bells. Enthusiasts take about three hours to toil through the 5040 permutations on seven bells; to do justice to eight bells would take a day. If you are an interrogator for the 'American-led coalition', given to playing continuous music to your captives to induce them to confess, you might consider an interesting variant. In the atonal twelve-tone music popularised by Schönberg, a composition is based on the twelve notes of the chromatic scale arranged in some order. There are 12! = 479001600 such 'tone rows' and it would take around 30 years to play them. After this, a new generation of interrogators could take over in the long task of fighting the war on terror.

Alternatively, Sudoku could provide a suitable life-long task. By developing these counting procedures, it has been found that there are 6670903752021072936960 patterns for 3 x 3 Sudoku: in solving a puzzle you are rejecting all but one of these possibilities. However, you might say some of these patterns are essentially the same. They differ only through the permutations of columns or rows, a rotation of the square, or a renaming of the symbols. The accepted figure for the number of *essentially different* patterns is 5472730538.

EASY: For $n \geq 5$, $n!$ always ends with a 0.
DIFFICULT: There are many zeroes at the end of 100! and 1000! – how many? (Hint: do not work these numbers out.)

Odd balls

A cube has complete symmetry between its six faces. This symmetry determines the properties of *dice*, with six equally likely outcomes, each having a probability of 1/6. In passing, it is worth noting that there are four other shapes with such complete symmetry: tetrahedron, octahedron, dodecahedron and icosahedron. But they lack the practical convenience of cubes. So that six-faced cube makes a natural link between the counting of possibilities, and the measuring of chances through the theory of probability.

The meaning of 'probability' and 'randomness' is not as simple as one might like. Sometimes probability is used as if equivalent to frequency or prevalence. That's absolutely true, as far as anyone knows, when randomness derives from a quantum-mechanical measurement. In the case of dice, however, randomness means only that the outcome depends in such a complicated way on how they are shaken and thrown that in practice it cannot be predicted. This, ultimately, depends upon the chaotic nature of physical systems, and the sensitive dependence of outcomes on initial conditions. So it has its limitations.

If dice-throwing was studied as intensively as golf swings, perhaps there would be champions in achieving double-sixes. (Conversely, if I played golf the results would be about as good as throwing dice.) An example on the borderline of randomness arises in the playing of roulette. This is intended by casino operators to be a game of pure luck, but in the 1980s a group of scientists noticed that because bets can be placed after the ball has been set rolling, it is actually possible to make a quick prediction and take advantage of it. They smuggled in a computer, and until thrown out of Las Vegas, made good money out of science.

A deeper problem is that sometimes the word 'probability' is used for describing a state of belief, rather than something observed and measured. This question of whether probability is objective or subjective goes back to its earliest days. It is natural to use this more subjective meaning in situations where there will only be one outcome, as yet unknown – for instance, the future climate of the planet. Controversy about the forecasts of the Intergovernmental Panel on Climate Change has so far not focused on its definitions of probabilities, but it is worth noting that this is a difficult question, taken very seriously by climate scientists. They have developed methods for calculating probabilities for outcomes, based on taking large ensembles of slightly different models, parallel Earths with slightly different assumptions. The IPCC also includes a more subjective element in its quantification of expert confidence in the conclusions reached. It is hard to see quite how these numbers can be tested by any actual measurement, or based on frequency.

It is a relief to turn from an area where probability is vitally important, but difficult to rationalise completely, to one which is generally foolish, but crystal clear. Probability has a perfect illustration in the casino, and for many people their most intense encounter with numbers comes in – well, the numbers game. And as it happens, the past decade has seen an explosion in the gambling business, with the British government promoting it as an economic lifeline for regions such as Manchester which were once proud centres of science and industry. (A parallel development of gaming and gambling has changed the agricultural and manufacturing America known to Constance Reid in the 1950s.)

In the British national lottery, the prizes depend on guessing which six numbered balls will be spewed out randomly (as far as one can judge) with much razzmatazz from an urn of 49. Neglecting all questions of bonus numbers and secondary prizes, we can ask how many possibilities there are for these six balls, and so the probability of guessing them correctly.

There are 49 possibilities for the first ball to emerge from the urn, then 48 for the second, 47 for the third; 46, 45, and 44 for the remaining three. This makes 49 x 48 x 47 x 46 x 45 x 44 possibilities. But this is overcounting: the order in which they emerge is irrelevant. There are 6! possible orders, so the complete number of possible sets of six balls is

$$\frac{49 \times 48 \times 47 \times 46 \times 45 \times 44}{6!}$$

This is written more neatly as

$$\frac{49!}{43! \times 6!}$$

and this expression tells a story: it shows a symmetry between the 43 and the 6. This is a basic yes-or-no symmetry inherent in the problem. When you fill in the lottery card, you may do it either by choosing the six to be marked, or the 43 not to be marked, and the result will be the same.

This number can be worked out as 13983816. Assuming these outcomes are equally likely — as they are unless the lottery is as corrupt as football — a punter buying one card and marking it with six guesses has a chance of one in 13983816 of winning the lottery prize.

If you play over and over again, the average value of the entry card will be the value of the prize, divided by 13983816. This is less than the price of the card, but of course people don't buy lottery tickets for their average value, but for the pleasure of knowing that there is a non-zero probability of their lives being transformed by wealth. (They can also derive satisfaction from contributing to the arts, sports and community projects funded from the lottery profits.)

The lottery promoter caters further to irrationality by publicising the numbers that have come up more often in the past. I do not know whether punters consider it a good idea to choose such numbers because they are lucky, or whether they believe such numbers should be avoided because 'by the law of averages' they are less likely to appear in the future. Both lines of thought are misguided: assuming that the balls are juggled properly, the past record has no influence on the future. If Mars or Mithras have a soft spot for you, and can nudge electrons to make the balls in the urn emerge in line with your ticket, then you are on to a good thing. Otherwise, there are no lucky or unlucky numbers. Your best bet is to pick numbers that you think other, more foolish people will avoid. Then in the unlikely event of winning, your prize will not be shared.

Codes and ciphers

As the lottery shows, combinatorial numbers based on just a few elements can easily reach astronomical proportions. Another example comes in solving anagrams for crossword puzzles. A fifteen-letter anagram can need up to $15! = 1307674368000$ possibilities to be considered. In Umberto Eco's novel *Foucault's Pendulum* there is a discussion of a 'giant computer' for such tasks. But you don't need a computer, still less a giant one, for crosswords. This is because you can break these numbers down as quickly as they are built up, using features of language to eliminate huge numbers of possibilities at once.

This is why code-breaking is possible. Simple cryptograms, based on substituting one letter for another, can be made in 26! different ways, which means that to break one you have to locate the correct solution out of 403291461126605635584000000 possibilities. This figure suggests an impossible task, but you can solve such puzzles with a message of 30 or so letters, by such observations as that E is the most common letter, QKZ is an unlikely pattern, and so on. In contrast, the four-digit PIN security used on mobile phones cannot be broken down, so that celebrity-stalking journalists are forced to go through 10000 possibilities to get at their targets unless they can make a lucky guess.

These ideas, and the number six in particular, have played a vital part in world history. The Enigma enciphering machine, as adapted for German military use during the mid-1930s, had three rotors which could be used in any order, giving six possibilities. Polish mathematicians developed a brilliant method for breaking the resulting ciphered messages, but this depended on there being only these six rotor-order choices. When in 1938 the Germans added two alternative rotors, the choices went up from 3 x 2 x 1 to 5 x 4 x 3, a tenfold increase. The Poles had to pass the problem on to the British, who soon developed code-breaking into a great strategic asset, but were likewise stretched by a further increase to 8 x 7 x 6. These numbers were critical: the code-breaking process was on a knife-edge of practicability. If there had been 10 x 9 x 8 rotor choices from the start, the Poles would never have been able to get going.

The general point lying behind these figures is that if combinatorial numbers have to be matched by the scale of physical resources (in the number of people, pieces of equipment, or time taken) then the task may well be impossible. The German designers of the military Enigma believed they had devised a scheme that would ensure just such security. The complexity of the machine was enhanced by an extra *plugboard* which performed a further scrambling operation. A typical plugboard setting consisted of a choice of ten pairs of letters. Since every such choice created a different code, the code-breaker had to work out what choice had been made. The number of ways of choosing ten pairs out of 26 letters is

$$\frac{26!}{10!\,6!\,2^{10}} = 150738274937250$$

which is indeed astronomical.

DEADLY: Why is this? If 13 pairs were used rather than 10, would the number of possibilities be increased?

Sometimes people refer to these enormous figures as 'the odds against breaking the Enigma'. This is highly misleading. One would not measure the odds against completing a crossword puzzle by looking at the probability of finding the solution by filling in letters at random. The structure of a crossword puzzle can be broken down, and so, it turned out, could the logic of the plugboard.

Alan Turing, who has already made more than one appearance, was the British mathematician who achieved this feat. He and another mathematician, Gordon Welchman, found a logical method of dealing with all plugboard choices at once. It was remarkably similar to the work of solving Sudoku by following a chain of logical deductions based on consistency requirements. Turing's method depended on knowing for certain the plaintext for around 25 letters of an Enigma message, very like the 25 or so numbers generally given as initial information for a Sudoku puzzle. You might wonder whether this means that Enigma-solving could be used as a newspaper puzzle. There are two reasons why not: one is that the technicality of the Enigma wirings, needed for the logical deductions, makes this infeasible as a pencil-and-paper problem. Another is that Turing's method for attacking an Enigma message required solving not just one Sudoku-like puzzle, but perhaps millions. (The scale of this number is given by $8 \times 7 \times 6 \times 26^3$, the number of possible rotor settings.) Still, the difficulty of Sudoku gives a good picture of the logical complexity of Turing's method. The fact that it could be implemented on a machine with 1940s technology, breaking a message in hours rather than weeks, was vital to the Second World War.

Deranged?

All animals and plants have evolved to cope with uncertainty. Using probability theory to enhance human capacities doesn't mean behaving recklessly: it means using reason and patience. Turing's probability theory gave vital extra tools for adding up faint clues to Enigma-enciphered messages: it was the appliance of science. But culture has a problem with probability, and displays highly ambivalent and contradictory attitudes. Entrepreneurs reward themselves as risk-takers, yet the markets are said to hate uncertainty. 'Security' is the word round which governments spin.

Ambivalence about gambling is particularly conspicuous. Being based on pure chance, involving no skill whatsoever, and pandering to irrational arguments, the lottery has emphasised the dubious value of reason and education. It is, for some reason, greatly encouraged. But other games require enormous skill, and are, equally unaccountably, considered disreputable. Poker, which has boomed through Internet gaming, is one such.

Although the hands dealt in any one game are a matter of chance, a serious player will in the long run succeed by rational application of the laws of probability. Online play eliminates the human element of poker-face nods and winks, although it introduces the problem of not being able to see whether other players are colluding. It is a model of what business is supposed to be like, rewarding patience, investment, rational strategic decisions and a cool nerve in the face of occasional loss. Cool students rightly find this more a rewarding form of income than stacking supermarket shelves or providing escort services.

Notwithstanding these observations, an English court adjudicated in January 2007 that as the cards are shuffled, poker is a game of luck and not of skill. This is a judgment which is tantamount to rejecting the 500 years of mathematical probability theory, not to mention the insurance and financial industries which depend on it. (One of the most striking developments of mathematics since the 1980s, now providing lucrative employment for many graduates, is in the 'stochastic theory' of mathematical finance. It is based entirely on concepts of random behaviour.) Meanwhile the American legislature effectively criminalised on-line poker, sending its London-based finances into a nosedive.

The law can be an ass, but the laws of probability are rational, and poker gives good examples of how they can be applied. A typical problem, found on a poker website, is: 'What is the probability that a five-card poker hand contains at least one ace?' To solve it, consider the chance of a no-ace hand. The calculation can be summarised like this: there is a probability of 48/52 that the first card is not an ace, then 47/51 that the second is not an ace, and so on for the remaining three cards. The probability of five not-ace cards is the product $(48 \times 47 \times 46 \times 45 \times 44)/(52 \times 51 \times 50 \times 49 \times 48) = 0.658842...$ so the chance of at least one ace is $0.341158...$

The analysis of problems like this is much assisted by the idea of *conditional* probabilities. Given that you have one ace, what is the probability that an opponent has two? Such calculations can be used to measure the value of *information*. Indeed, Turing's deepest contribution to the code-breaking work lay in showing how to define this measure by objective, numerical procedures, and so turn it from an art into a mathematical science.

Valuable mathematics can be found even in the much

simpler game of Snap. Two packs of 52 cards are shuffled and each dealt out one at a time simultaneously: a snap is when the two cards coincide. This is a game for children (such as I played with my granny while Constance Reid was writing her book in the 1950s) but let's be grown-up and pretend we are betting on the number of snaps that occur during the stretch of 52 cards. If this was a television programme I could dramatise this tame game as Strip Snap, or as some globe-trotting tale about defusing bombs or Russian roulette. More seriously, you might see it as related to real problems like matching segments of DNA.

In fact, suppose we are placing bets on there being no snap at all in the 52-card run. What odds should you accept? Is a snap virtually certain, or most unlikely?

This question is easier to investigate for packs not of 52 but of fewer cards. If there are only two cards, then there are only two possibilities: we can take the first pack to be in the order AB, and the second one is either AB or BA. So the probability of no-snap is $1/2$. If there are only three cards, then we can take the first pack to be in order ABC, and there are six possible orderings for the second pack. Of these, only BCA and CAB have no snap. So in this case the probability of no-snap is $1/3$. For four cards, the no-snap orderings are the nine BADC, BCDA, BDAC, CADB, CDBA, CDAB, DABC, DCAB, DCBA out of the 24. The probability of no-snap is $9/24 = 3/8 = 37.5\%$.

For five cards, there are 44 no-snap possibilities out of 120 (36.67%), and for six cards there are 265 out of 720 (36.81%). If you try to write these down you will realise that you need a more systematic way of counting them. This is not so easy and I'll give the answers for seven, eight and nine cards in a way that shows the pattern:

$7! \times (\frac{1}{2!} - \frac{1}{3!} + \frac{1}{4!} - \frac{1}{5!} + \frac{1}{6!} - \frac{1}{7!}) = 1854$ out of 5040 (36.785714%).

$8! \times (\frac{1}{2!} - \frac{1}{3!} + \frac{1}{4!} - \frac{1}{5!} + \frac{1}{6!} - \frac{1}{7!} + \frac{1}{8!}) = 14833$ out of 40320 (36.788194%).

$9! \times (\frac{1}{2!} - \frac{1}{3!} + \frac{1}{4!} - \frac{1}{5!} + \frac{1}{6!} - \frac{1}{7!} + \frac{1}{8!} - \frac{1}{9!}) = 133496$ out of 362880 (36.787919%).

The pattern shows that it hardly makes any difference to the answer whether there are 52 cards or only five. The proportion of non-snap orderings gets closer and closer to the number given by

$$\frac{1}{2!} - \frac{1}{3!} + \frac{1}{4!} - \frac{1}{5!} + \frac{1}{6!} - \frac{1}{7!} + \frac{1}{8!} - \frac{1}{9!} \dots$$

which to twelve decimal places is .367879441171...

There is thus a probability of about 37% of no-snap, 63% of at least one snap. The 37% figure is not an amazing number, being a middle-of-the-road figure like the proportion of people believing in the imminence of Armageddon, the dinosaurs in Noah's Ark, and abductions by aliens. But this is one of the most fundamental numbers in mathematics. This limiting proportion of no-snap orderings (called 'derangements') has a special name. Actually the name is given to its reciprocal, the odds against no-snap in an infinite run. This is called e. So that probability of about 37% is called e^{-1}, while e itself is the number 2.718281828459...

There is another fact about coincidences which is generally found astonishing: in a group of 23 people, it is more likely than not that two of them share the same birthday. But if you know about e, this should not come as such a surprise. Here is a rough (not a precise) argument: in a group of 28 people there are 378 pairs of people whom you can introduce and ask if they share a birthday. In each case there is a chance of $1/365$ that they do, so your knowledge of e leads you to expect a better than 63% chance of hearing the words 'Oh that's incredible!' In a group of 23 the chance is lower, but it is not surprising that it is just over 50%.

High on e

We can re-express e^{-1} in a more elegant way which brings out a pattern involving all the integers:

$$e^{-1} = \frac{1}{0!} - \frac{1}{1!} + \frac{1}{2!} - \frac{1}{3!} + \frac{1}{4!} - \frac{1}{5!} + \frac{1}{6!} - \frac{1}{7!} + \frac{1}{8!} - \frac{1}{9!} \cdots$$

This makes sense if we define $1! = 1$ and $0! = 1$, so that the two extra terms just cancel each other. Remember that we have used the pattern $4! = 4 \times 3!$, $3! = 3 \times 2!$, so we can fix $1!$ and $0!$ by consistency with $2! = 2 \times 1!$ and $1! = 1 \times 0!$.

The beauty of this is that by changing all the minus signs to plus signs we get another true formula:

$$e = \frac{1}{0!} + \frac{1}{1!} + \frac{1}{2!} + \frac{1}{3!} + \frac{1}{4!} + \frac{1}{5!} + \frac{1}{6!} + \frac{1}{7!} + \frac{1}{8!} + \frac{1}{9!} \cdots$$

The number e is unique in having this simple relationship between powers and permutations. The duality between + and – in these formulas should suggest a connection with the symmetries of Chapter 2. The two formulas can be rolled into one by the fuller statement that

$$e^r = \frac{1}{0!} + \frac{r}{1!} + \frac{r^2}{2!} + \frac{r^3}{3!} + \frac{r^4}{4!} + \frac{r^5}{5!} + \frac{r^6}{6!} + \frac{r^7}{7!} + \frac{r^8}{8!} + \frac{r^9}{9!} \cdots$$

This formula comes in naturally when we extend the scope of investigation away from simple Snap.

To do this, we return to the question of the lottery. Remember that the chance of any player winning is 1 in 13983816. Suppose first that there are exactly 13983816 players taking part. You would be right to think that the *average* number of winners is then 1. But in any draw, there might be 1, 2, 3... winners – or none at all. If there are no winners, the lottery prize money is 'rolled over' to the next draw. What is the chance of this happening? For any individual player the probability of not winning is 13983815/13983816. Assuming that the 13983816 players make their guesses independently, the probability that they all fail to guess correctly is $(13983815/13983816)^{13983816}$. So this is the probability of a rollover.

It turns out that this number is (almost exactly) e^{-1} again. Rather than try to work it out, consider the sequence of numbers $(1/2)^2$, $(2/3)^3$, $(3/4)^4$, $(4/5)^5$, $(5/6)^6$, $(6/7)^7$... Your calculator will give

.25, .296296, .316406, .327680, .334898, .339917, .343609, .346439, .348678...

You can see these numbers converging, though much more slowly than the no-snap probabilities, to e^{-1}. The more general formula comes in like this: if $r \times 13983816$ players are participating, then the probability of a rollover is (almost exactly) e^{-r}.

A closely related problem is this: spammers and phishers rely on there being a sucker born every minute. Suppose this is true: not that they are born regularly every 60 seconds, but randomly with an average rate of one a minute. What is the probability of a minute elapsing without a sucker being born? It is e^{-1}. What is the probability of r minutes elapsing without a birth? It is e^{-r}.

Another classic example of e lies in the vicious circle associated with compound interest. If tempted to consolidate your credit with the Lone Shark Finance Co., as advertised on daytime television, you should watch out. Suppose the rate is 100% per annum, with the small print saying that there shall be no lower limit on the time period on which compound interest is computed. How much interest can Mr Shark demand on £1000 after a year?

At first sight you might think it was £1000. But Mr Shark's small print has got teeth. He could argue that the interest should be added daily, and that each day the debt increases by $(1 + 1/365)$. At the end of the year the debt would then be £1000 x $(1 + 1/365)^{365}$. At this point you might fear that by demanding hourly, minutely, secondly, nanosecondly interest the debt could be made as high as Mr Shark wanted. But e comes to your rescue – the limit as the calculation gets finer and finer is just £1000 x e. If Mr Shark's interest rate is not 100% but r x 100%, then the limit is £1000 x e^r.

Growth and oscillation, e and π

The last example, of vicious compound interest, shows

how the number *e* is related to *growth*. This means it is also intimately connected, by the duality of growth and oscillation we met in Chapter 2, to the concept of cycles and circular motion. The full understanding of this relationship requires the full story of the *complex numbers*, but it is possible to pick out some illustrations without giving this complete picture. In particular, *e* is related to the more famous number π.

Using the complex numbers, a single, famous equation defines the relationship:

$$e^{i\pi} = -1.$$

Here *i* means exactly the same as the complex-number pair (0, 1), described in Chapter 2 as the square root of –1. Alternatively, without using complex numbers, this equation is exactly equivalent to asserting the two properties of π:

$$\frac{\pi^1}{1!} - \frac{\pi^3}{3!} + \frac{\pi^5}{5!} - \frac{\pi^7}{7!} + \frac{\pi^9}{9!} - \frac{\pi^{1}}{11!} \ldots = 0$$

$$\frac{\pi^0}{0!} - \frac{\pi^2}{2!} + \frac{\pi^4}{4!} - \frac{\pi^6}{6!} + \frac{\pi^8}{8!} - \frac{\pi^{10}}{10!} \ldots = -1$$

Rather than try to go further with complex numbers, I'll point out the way that these expressions, involving powers of π and factorials, are related to the ideas about many-dimensional spaces which have appeared in the earlier chapters.

You may be familiar with the formula that says that π is the area enclosed by a circle with unit radius. But the geometrical meaning of π is not confined to circles. It turns out that $\pi^2/2!$ is the hypervolume enclosed by a unit-radius 3-sphere in four dimensions, as defined in Chapter 4. The same pattern continues, so that the hypervolume enclosed by a unit $(n-1)$-sphere in n dimensions is:

$$\frac{\pi^{n/2}}{(n/2)!}$$

This formula looks as if it makes no sense for odd values of n. But for $n = 3$, it gives the correct answer for the volume of a unit sphere $(4\pi/3)$, if $(3/2)!$ is defined to be $3/4 \times \sqrt{\pi}$.

This turns out to be a completely consistent definition: in fact $(-1/2)! = \sqrt{\pi}$, so that $(1/2)! = 1/2 \times (-1/2)! = 1/2 \times \sqrt{\pi}$, and so indeed $(3/2)! = 3/2 \times (1/2)! = 3/4 \times \sqrt{\pi}$. The rest follow on the same pattern, giving a correct value for all the hypervolumes.

What we have done here is to extend the meaning of the ! symbol for factorials, so that it is no longer defined by the number of ways of permuting a number of objects. This is very similar to the way that the concept of a power was extended in Chapter 4 from positive integers to zero, negative integers and fractions. It gives a beautiful example of how mathematics grows in surprising directions, as more subtle inter-relationships are discovered. A definition of $z!$ can likewise be given for almost all values of z, in such a way that the relationship $z! = z \times (z-1)!$ always holds good. This is true for complex-number values as well. One must say 'almost all values' because z cannot be a negative integer.

Thus the number $\sqrt{\pi}$ opens the way to the measurement of hypervolumes enclosed by hyperspheres. That might seem at first sight to be of no use whatsoever, but many-dimensional spaces can be thought of as game-strategy spaces, stock-control spaces, fashion-following spaces, and generally anything from the complexity of life. A hyper-sphere is defined by extending Pythagoras's metric into many dimensions, so the hypervolumes are all to do with sums of squares. If you are assessing how well customer demand has fitted your stock-ordering policy, then this adding up of squares is a good way to do it.

Measuring the hypervolume of a hypersphere can be thought of as measuring the probability of a random point lying inside it. So hypervolumes are naturally connected with the probability of complex events happening: statisticians use such n-dimensional spaces (called 'degrees of freedom') all the time. The exact mathematical result is this: suppose you have random numbers evenly spread over the range from –1 to 1. Take n such numbers, and add up their squares. The probability that this sum is less than 1 is

$$\frac{1}{2^n} \frac{\pi^{n/2}}{(n/2)!}$$

DIFFICULT: Work out the values for n = 2, 3, 4, 5... Why do they decrease as n increases? How are the probabilities related to this picture?

A random walk

There is an even more natural setting in which $\sqrt{\pi}$ comes into life's choices and chances. Imagine taking a *random walk* along a street: for each step you toss a coin to decide whether to go forward or backward. This you can take as a metaphor for the up-and-down nature of life, or as something that enters into almost any situation where many random changes are combined, for instance in the behaviour of gases, the rounding errors in a spreadsheet, or the spread of an epidemic. If you take n such random steps, it is just possible, but very unlikely, that you will go n steps in one direction. It is much more likely that you will end up near where you started. How likely? How near?

There is a useful way of visualising the possible outcomes: the Pascal Triangle. This is easily written down because every number is the sum of the two above it.

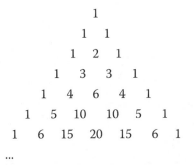

The nth row now gives the distribution of probabilities for the various possible outcomes of your n steps. Another way to visualise this triangle is as a pinball machine, with balls falling from the top and knocked randomly to left or right as they cascade downwards. They are more likely to end up in the middle than at the edge, and the numbers in the triangle give an accurate account of the distribution.

EASY: The rows add up to the powers of 2. Why is this? Find diagonals on which the entries sum to the Fibonacci numbers.

The triangle also has a direct connection with the lottery calculation we did earlier. This is because the $(r + 1)$th number in the nth row gives exactly the number of ways of selecting r things out of n. For instance, 13983816 would appear as the seventh number in the 49th row. I do not recommend checking this but the 6 in the fourth row gives the number of ways of choosing two out of four, and you can check there are six: (12) (13) (14) (23) (24) and (34).

As you look at larger and larger values of n, the shape made by the numbers, like the stack of balls at the bottom of a huge pinball machine, settles down to a very specific shape – often called 'the bell curve'. Gauss discovered the universal character of this curve, which makes it the foundation of all serious statistical analysis. The formula for it involves e and π, and here is one fact which illustrates it: after an even number of steps, there is a definite probability of your random walk being back exactly where it started. (This is equivalent to the height of the central point of the pinball machine stack.) What is that probability? After n steps, it is very close to

$$\frac{\sqrt{2}}{\sqrt{\pi} \times n}$$

The point is that π should not be thought of just as the measurement of a circle. It is a deep structure embedded in the numbers, bound up with growth, probability, permutations, complex numbers, and many-dimensional geometry.

These subtle relationships lead to means by which it can be calculated.

Squaring the circle

The number π has always got a good press, and gets a boost from people who take great pleasure in finding and even memorising huge numbers of decimal digits, starting with 3.14159265358979323846264338327950288419 7... But a question that is not so often asked is that of how these decimal digits are actually found. Although it is vaguely known that mathematicians can 'calculate it', how would you even begin to do the calculation? Certainly it is not obtained by measuring circles! Measurements can only have limited accuracy. Besides, general relativity tells us that space-time is everywhere curved, so the concept of a perfect circle is inconsistent with the actual physical world. The expression for e in terms of factorials gives a picture of the kind of infinite formula that must be found. But π doesn't have such a convenient expression, and mathematicians from Archimedes onwards have been led into deeper water in trying to find practical formulas.

The startingpoint is that π is a *logarithm*. In fact, π is the imaginary part of the logarithm of (-1) to base e, as follows from the complex-number relation $e^{i\pi} = -1$.

The logarithm is the inverse of an exponential, and there are ways of inverting the formula for e^r, as given earlier, to obtain formulas for logarithms. These in turn lead to formulas for π. They give striking pictures of how π is interwoven with the integers, even though they are not very practical for calculations.

First comes a classic formula for squaring the circle:

$$\frac{\pi}{4} = 1 - \frac{1}{3} + \frac{1}{5} - \frac{1}{7} + \frac{1}{9} - \frac{1}{11} \cdots$$

But it has a more advanced version:

$$\frac{\pi^3}{16 \times 2!} = 1 - \frac{1}{3^3} + \frac{1}{5^3} - \frac{1}{7^3} + \frac{1}{9^3} - \frac{1}{11^3} \cdots$$

There is a similar series for every odd power. But new strange factors come in:

$$\frac{5 \times \pi^5}{64 \times 4!} = 1 - \frac{1}{3^5} + \frac{1}{5^5} - \frac{1}{7^5} + \frac{1}{9^5} - \frac{1}{11^5} \cdots$$

$$\frac{61 \times \pi^7}{256 \times 6!} = 1 - \frac{1}{3^7} + \frac{1}{5^7} - \frac{1}{7^7} + \frac{1}{9^7} - \frac{1}{11^7} \cdots$$

$$\frac{1385 \times \pi^9}{1024 \times 8!} = 1 - \frac{1}{3^9} + \frac{1}{5^9} - \frac{1}{7^9} + \frac{1}{9^9} - \frac{1}{11^9} \cdots$$

$$\frac{50521 \times \pi^{11}}{4096 \times 10!} = 1 - \frac{1}{3^{11}} + \frac{1}{5^{11}} - \frac{1}{7^{11}} + \frac{1}{9^{11}} - \frac{1}{11^{11}} \cdots$$

The numbers 1, 1, 5, 61, 1385, 50521, 2702765, 199360981, 19391512145, 2404879675441... are the *Euler numbers*. They make a beautiful illustration of the interconnectedness of π with permutations, which calls for another excursion into the world of music to explain. Most well-known tunes can be identified simply by whether successive notes of the melody rise or fall. Now, in ringing the changes on six bells, sometimes the tune goes down-up-down-up-down, as for instance in the sequences:

If you count the number of tunes with this property you will find just that Euler number 61. For eight bells there are 1385, for ten bells 50521, and so on.

There is a another relationship between the Euler numbers and π.

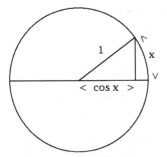

Take a circle of unit radius. Mark off a part of the circle of curved length x. It defines a straight length called cos x, as shown. As x increases from 0, cos x decreases from 1. The exact way that this happens is given by the formula:

$$\cos x = \frac{1}{0!} - \frac{x^2}{2!} + \frac{x^4}{4!} - \frac{x^6}{6!} + \frac{x^8}{8!} - \frac{x^{10}}{10!} \cdots$$

which you may notice is closely related to the formula for e^x, a further aspect of the duality of growth and oscillation. If we turn this upside down, the Euler numbers appear:

$$\frac{1}{\cos x} = \frac{1}{0!} + \frac{x^2}{2!} + \frac{5x^4}{4!} + \frac{61x^6}{6!} + \frac{1385x^8}{8!} + \frac{50521x^{10}}{10!} \cdots$$

You may have noticed that after the initial 1, the Euler numbers end alternately in 1 and 5. This pattern continues for ever and the final two figures also repeat in a cycle of length 10. Seeing why this is true is a more than SUPER FIENDISH puzzle. It is left as a further tantalising picture of how the patterns of primes, powers and permutations are tangled up with the number π.

Probability and primes

A closely related formula found by Euler, bringing us back to the number six, is

$$\frac{\pi^2}{6} = 1 + \frac{1}{2^2} + \frac{1}{3^2} + \frac{1}{4^2} + \frac{1}{5^2} + \frac{1}{6^2}\ldots$$

This formula is the starting point for another, quite unexpected, connection. It is related to the *prime numbers*. Seeing this connection needs one of the slickest pieces of algebra in all mathematics, again something found by Euler. It is as perfect as $1 + 2 + 3 = 1 \times 2 \times 3$, turning addition into multiplication. In fact, this is just what we will do.

The first step transforms the infinite sum into an infinite multiplication:

$$\left(1 + \frac{1}{2^2} + \frac{1}{4^2} + \frac{1}{8^2} + \frac{1}{16^2}\ldots\right) \times \left(1 + \frac{1}{3^2} + \frac{1}{9^2} + \frac{1}{27^2} + \frac{1}{81^2}\ldots\right)$$
$$\times \left(1 + \frac{1}{5^2} + \frac{1}{25^2} + \frac{1}{125^2} + \frac{1}{625^2}\ldots\right) \times \left(1 + \frac{1}{7^2} + \frac{1}{49^2} + \frac{1}{343^2} + \frac{1}{2401^2}\ldots\right) \times\ldots$$

The truth of this statement depends on the *Fundamental Theorem of Arithmetic* we met in Chapter 1. To see how it works, take a particular term in the infinite sum, say $1/84^2$. It can be written as

$$\frac{1}{4^2} \times \frac{1}{3^2} \times 1 \times \frac{1}{7^2} \times 1 \times 1\ldots$$

and this product corresponds to just one of the terms that results from multiplying out all the factors in the infinite multiplication. The same applies to every integer.

The second step is to simplify these factors using a formula for infinite additions, borrowed from Chapter 9:

$$1 + \frac{1}{p^2} + \frac{1}{p^4} + \frac{1}{p^6} + \frac{1}{p^8} + \ldots = \left(1 - \frac{1}{p^2}\right)^{-1}$$

The result is best appreciated when turned upside down:

$$\frac{6}{\pi^2} = \left(1 - \frac{1}{2^2}\right) \times \left(1 - \frac{1}{3^2}\right) \times \left(1 - \frac{1}{5^2}\right) \times \left(1 - \frac{1}{7^2}\right) \times \left(1 - \frac{1}{11^2}\right) \ldots$$
$$= 3/4 \times 8/9 \times 24/25 \times 120/121 \times \ldots$$

Each of the factors can be interpreted as a probability. Imagine that we pick two integers at random. The first factor, 3/4, is the probability that two numbers are *not both even*. The second factor, 8/9, is the probability that they are not both divisible by 3. The third factor is the probability that they are not both divisible by 5, and so on. The product of all the factors is the probability that *two randomly chosen integers have no prime in common*. Equivalently, it is the probability that the *highest common factor* of two integers is 1. This probability is .6079..., a curiously mediocre answer for such an extraordinary question.

You may wonder how probability can come into statements about primes and factors: numbers don't have any uncertainty in their factors. The answer is that the randomness goes into the method for picking the numbers, not into the properties of the numbers you pick. It is worth spelling out what this means. First, take the numbers 1 to 9, and assume you have a method for picking one of them randomly and with equal probabilities. Choose a pair of numbers by using this method twice over. Of the 81 possible pairs, just 55 of them have a highest common factor of 1 – which gives a 68% probability. The statement is that if you do the same thing for the numbers 1 to n, the proportion gets closer and closer to $6/\pi^2$ as n gets larger and larger.

There is similar argument for the probability that k integers have no prime factor in common, namely the reciprocal of

$$1 + \frac{1}{2^k} + \frac{1}{3^k} + \frac{1}{4^k} + \frac{1}{5^k} + \frac{1}{6^k} \cdots$$

When k is even, this can be summed to a term involving π^k and one of the *Bernoulli numbers*. These are closely connected with the Euler numbers, to the geometry of circles, and to the counting of up-and-down tunes (with an odd number of notes). The next step is to ask what happens if k is something other than an even number.

Beyond growth and oscillation

The question is closely connected with the distribution of prime numbers, which gradually become rarer and rarer. Gauss had already seen that the density of prime numbers near n behaves like the logarithm (to base e) of n. This means that for 100-figure numbers, about 1 in 230 is prime; for 200-figure numbers about 1 in 460 is prime, and so on. Other mathematicians had begun to explain this pattern, but it was Riemann who made a great advance in 1859. His method was analogous to the extension of the factorial to non-integer values. Riemann took the probability that k integers have no prime factor in common. He extended k not just to odd integers, or half-integers, but to all complex numbers. This defined the *Riemann zeta-function*. Using the duality of growth and oscillation, Riemann could then express the thinning out of the primes in terms of a series of waves. The mathematician Marcus du Sautoy has called these waves 'the music of the primes'. Riemann saw that if the music takes a very particular form, much could be learnt about the infinite sequence of primes, and the exact way they thin out. This conjecture about the waves is the *Riemann hypothesis*.

The truth of the Riemann hypothesis is still a conjecture, and the subject of another Millennium Prize. In fact it has been a prominent problem for 100 years, and is much the oldest to attract the million-dollar bounty now. It seems to mark a frontier of the landscape that is carved out by starting with e and π, and which embodies the connections between growth, rotation, natural numbers and primes.

In a sense it is surprising that it remains unsettled because such completely new insights into numbers have emerged since Riemann's time. Amongst all the contributors, one utterly magic figure is that of Srinavasa Ramanujan. He wrote to Hardy out of the blue from Madras in 1912–13 with a gallery of completely new formulas, presented as *faits accomplis*, since he rarely gave proofs. For Hardy it was rather like Turner getting a preview of the Turner Prize. Ramanujan had an extraordinary intuition not just for numbers, but for relationships which emerge on going beyond the landscape of the exponential function, and into the mountain range formed by the *elliptic functions*. This is territory which, 100 later, is central to mathematical discovery.

As a glimpse of new relationships, he could see why $e^{\pi\sqrt{58}}$ is almost an integer, being $396^4 - 104.000000177...$ On the basis of this, Ramanujan found a formula for π, which gives a far quicker way of calculating it:

$$\frac{9801}{\sqrt{8} \times \pi} = 1103 + \frac{4! \times (1103 + 26390)}{396^4} + \frac{8! \times (1103 + 2 \times 26390)}{(2!)^4 \times 396^8} + \frac{12! \times (1103 + 3 \times 26390)}{(3!)^4 \times 396^{12}} \cdots$$

Ramanujan noted this in about 1913, but it was proved only in the 1980s. Similar formulas, slightly more complicated, are even better and are used for calculating huge numbers of decimal places. A splendid book by David Blatner, *The Joy of Pi*, prints a million decimal places as found by these means. There are many such formulas, but they always agree. It is a striking fact that the consistency of mathematical truth, of what Hardy said 'is so', can be used as a powerful practical principle: the agreement is used to check that computers are working properly.

What is it that makes the elliptic functions go beyond the exponential? As a rough picture, the difference is that they are based not on the geometry of planes and spheres, but of surfaces like the torus, with *holes*. They are vital to string theorists, for whom such surfaces are the starting point. They are also related to the advanced structures used by Andrew Wiles to prove Fermat's great conjecture. A bigger surprise, at the end of the twentieth century, was the discovery of a close connection with the algebra of the finite simple groups, including the Monster group. The discoveries so far must be only the first steps in exploring this territory, the analogues of what Bernoulli and Euler did 300 years ago. A reincarnation of Ramanujan would find enormous scope for exercising magical insight.

Lottery of life: tall stories

Most things that happen are highly unlikely. If you play bridge, it is almost certain that each new deal sets up a game which has never been seen before in the history of the world, and will never be seen again. There are $(52!)/(13!)^4$ possible bridge hands, which is more than 10^{28}, and it would take ten billion bridge parties, each playing a million trillion games, to work through them.

There are no likely lads or lasses, since the sexual splicings of DNA are like card deals. Everyone has been dealt a highly unlikely hand. Richard Dawkins, in *The Ancestor's Tale,* points out how few of us are alive compared with those who could have been. Dawkins's general description of evolution is that of climbing Mount Improbable, his message being that the improbable is not as impossible as it looks at first, or perhaps noughth, sight.

Snap shows how although an individual coincidence may be surprising, to have a coincidence *somewhere* is not surprising at all. A similar argument may be made about coincidence in ordinary life: there are so many surprising things that *might* happen in a day, that it is not remarkable at all to have one surprise. The origin of self-reproducing DNA precursor molecules may have been a very improbable chemical coincidence, unlikely to be reproduced in a laboratory, but if there were enough opportunities for the coincidence to occur then it is no miracle. Astrological predictions will be right now and then, and if you only take note of the successes, you will be impressed by them.

Unlikely events give rise to irrational arguments. Suppose Adam and Steve both buy a lottery ticket, and forget which is which. One of the tickets turns out to win the jackpot. Adam says: 'Steve, the chance of your winning the lottery was one in thirteen million, it can't be you.' Steve says: 'Adam, the chance of your winning the lottery was one in thirteen million, it can't be you.' The right answer is that the *conditional* probability of either of them winning, given that one of them did win, and in the absence of further information, is 1/2. They should share the money equally. This may seem obvious, but it is not. In 1999 a mother was convicted of the murder of her child by a jury bamboozled by the 'expert' argument that there were enormous odds against it having been an accidental 'cot death'. The subsequent controversy focused on the enormity of the odds given by the expert, which were 73 million to one, as opposed to the odds given by other experts of a few thousand to one. But what the jury needed was the *ratio* of this unlikelihood, whatever it was, to the unlikelihood of the alternative explanation, namely murder. Without a comparison with this figure (and the probability of a mother murdering her children is certainly small) any judgment is as fallacious as the self-serving arguments given by Adam and Steve.

A special case arises if one of the probabilities is actually zero. Examination of the ratio then boils down to Sherlock Holmes's dictum: when the impossible has been eliminated, whatever remains, however improbable, must be the truth. But the courtroom has not yet caught up with conditional probability, even though important legal cases hang on the probabilities of DNA matching, and other such forensic evidence.

Another example of the dubious use of probability lies in the expression 'the balance of probabilities', which is used to decide civil cases, but which is also urged as a means of allowing more scope for detentions without trial. It is supposed to mean that a probability of more than 1/2 decides the case. Suppose that the cases coming before a court range evenly from open-and-shut cases with probabilities of 0% or 100%, to 'nothing-to-choose-between-them' cases, with probabilities of 50% either way. Then one quarter of the decisions made will be wrong. If, more realistically, the cases occur mostly in the difficult centre ground, then nearer to half the decisions will be wrong, and it would be a lot cheaper to toss a coin for it. I doubt whether this statement of the outcome would command the same respect for the majesty of the law, as does the resounding verbal expression of 'balance'. But I suspect that few legal practitioners would think of putting their principle into cold numerical terms.

In contrast, exact and scientific accounts of probability are distrusted or laughed out of court. A notable case arose in a paper in the medical journal, the *Lancet,* in October 2006. The authors reported that 'We estimate that between March 18, 2003, and June, 2006, an additional 654965

(392979-942636) Iraqis have died above what would have been expected on the basis of the pre-invasion crude mortality rate as a consequence of the coalition invasion.' The range from 392979 to 942636 is a '95% confidence interval', fully justified by mathematical probability. But that theory, depending as it does on the work of Gauss, on hyperspheres, e and the square root of π, can readily be discounted as 'extrapolation' and 'academic'. It is of course true, as with any kind of random sampling survey, that if the sample had been biased in favour of locations with heavy death rates, then the 95% confidence in the result would be unjustified. Some critics alleged just this, although the authors of the survey offered detailed defences of their fieldwork. But the argument against believing the results on the basis of it being merely a projection, or merely academic, is essentially an argument against the validity of mathematical or scientific knowledge.

Probability tells you what to expect from a fair lottery, the science of statistics looks at the outcomes and asks how sure we can be that the lottery is fair. Statistics, in the grown-up sense of the word, does not mean the making of lists of figures, nor damned lies, nor proving anything with certainty, but making the best efforts at the rational deduction of cause from effect. Those efforts may err because of faulty assumptions, but at least mathematics makes those assumptions explicit, so that they can be identified and corrected. Even this achievement is highly worthwhile, because people generally adhere to their *a priori* beliefs, and accept or reject evidence according to how well it fits in. Fortunately for the

Anglo-American powers, the authorities of Nazi Germany took this approach to the Enigma.

Code-breaking in the Second World War gave a superb example of deducing the message from the scrambled signals. Science means breaking the code of Nature. But it's hard to decide whether its secret lies in Six, or in —

7

The Wonder of it All

When Alice was in the Looking-Glass land, Tweedledee recited a poem for her:

> The Walrus and the Carpenter
>   Were walking close at hand;
> They wept like anything to see
>   Such quantities of sand:
> 'If this were only cleared away'
>   They said, it WOULD be grand!'
>
> 'If seven maids with seven mops
>   Swept it for half a year,
> Do you suppose,' the Walrus said,
>   'That they could get it clear?'
> 'I doubt it,' said the Carpenter,
>   And shed a bitter tear.

Seven in German is *sieben,* the same word as for a sieve. Seven needs sifting and sorting out. If you need seven things from the shops, it is wise to make a list. 'Seventhly', it is said, is the most depressing word in oratory. Seven is one over the top, a bridge too far. The heptagon cannot be constructed by knowing √7. The seventh harmonic is the poorest fit to the equal temperament scale, and the natural horn's seventh was always problematic in classical music. The transition to romantic sound, with the whole orchestra singing as a single voice, needed the technological horn of the 1820s. In Vaughan Williams's Third Symphony, a subtle mourning for the First World War, the horn is instructed at one point to play its natural seventh harmonic as an unintegrated, unresolved, unsmoothed sadness.

Seven is also a celebrity, with many fans: vices, virtues, league boots, year itches, and of course dwarves. My favourite celebrities are the legendary Seven Sleepers who slumbered in a cave through the whole of the fascinating fourth century. They woke up to find the world had given up all the stories, symbols and rites of what the Romans called *religio.* Unlike Christiane Kerner in the wonderful *Goodbye Lenin,* who underwent a shorter coma, these troglodytic Christians were delighted by the turn of events.

But Seven is not usually so easily satisfied. Seven is the awkward customer who demands one over the odds, and comes back with the complaint that what they bought yesterday doesn't fit after all. Seven is the number of Nature, which has so far refused to be cleared up.

Sixes and sevens

A little noted aspect of the Seven Sleepers' time travel is that they dropped off in the old Roman world of the Ides of March and all that, and awoke on what was perhaps a Monday, or a Thursday morning. Constantine, in establishing Christianity as politically correct, had also made official the days of the week, with the *dies Solis* as its holiday. Astrology is the key to that cycle of seven names, which apparently have been found in Pompeii, and so known to be popular long before they were made official. An odd fact is that the old gods survived in them, despite the Church's efforts to eradicate them, and so live on in most European languages today. It is a fair guess that the Babylonian weeks, from which the subsequent systems seem to spring, evolved because seven days roughly mark a phase of the Moon and of the menstrual cycle. The pagan planets, translated into Latin or Germanic languages, gave the cycle of Sun, Moon, Mars, Mercury, Jupiter, Venus and Saturn, and now dictate the cycle from Sunday to Saturday. Indeed, their modulo-7 arithmetic dictates the frenetic work-week and market-trading spasms of the entire globalised world.

Astrology, like magic, is a first attempt, or at least a noughth attempt, at science. Given the splendour and drama of the night sky of antiquity (compared with the tiny world of streetlamps, shop windows and mobile phones now available to megalopolitan youth), a thesis of celestial involvement in the minutiae of human doings is a perfectly reasonable noughth impression of the cosmos. More thoughtful first impressions show a tendency to find not Seven-ness but Six-ness in the basis of the cosmos, perhaps from the 360-and-a-bit days of the year, and the perfection of $1 + 2 + 3 = 1 \times 2 \times 3$. The Mayans' 144000-day units, the Hindu age of 432000 years, and the great age of $622080000000000 = 6^5 \times 80$ billion years are examples. Six occurs in the twelve signs of the Zodiac, the calendar months, and the Babylonian 60-based numbers from which we inherit minutes and seconds of angle, and with them the minutes and seconds of time.

But 365 is that tiresome bit bigger than 360, which spoils the perfect party. The year is not even exactly 365 days, requiring the Gregorian fix of 146097 days in 400 years so as to keep the western Christmas where it was at the time of Constantine, a few days after the solstice. (The Orthodox feast will slowly migrate into the spring, a remote consequence of his division of the Empire.) The lunar cycle pushes in its non-fitting sevens, but is itself not quite 28 days, whichever way you measure it. These baffling ratios confronted early cultures with a puzzle of a truly TRICKY standard. Finding integers to make a solar calendar is difficult enough, like the problem of fitting $\sqrt{2}$, but to find a calendar that suits both Sun and Moon is as difficult as reconciling the inconsistent harmonic ratios of music.

As it happens, nineteen plays a role in the sacred calendar as it does in music, though for some reason this has never given it any of the glamour of Seven. (Nineteen has a famous song, but for a completely different reason.) The Sun-Moon system almost fits over nineteen years, during which the moon orbits 235 times. This coincidence determines the date of the Christian Easter, as formulated in the sixth century. The Byzantine monk Dionysius Exiguus adopted the period of 19 x 7 x 4 = 532 years as the basis of the Christian calendar, and set out a scheme of modulo-19 arithmetic for using it. This was a rare spark in the European dark age of number, while so much more important things happened in India and Persia.

Fixing the date of Easter was apparently far more important than finding historical evidence for the miracle it marks, and a similar remark could be made for the Nativity. It's worth noting that AD1 was fixed (more or less) by Dionysius Exiguus some five hundred years after the supposed events. The gospels themselves indicate that fulfilment of prophecy counted far more than historical details, and that mathematical pattern added extra value. Stretching a point, the three wise men are the nearest thing to scientists: as following a star is physically impossible, the story must refer to the supposed power of astrological prediction. A modern-day equivalent might be a supercomputer prediction by three top intelligence chiefs. Perhaps the only reference to number factorisation by the evangelists relates Seven-ness to the tricky question of Jesus's Y chromosome. Their genealogies give two (different) descents from King David in fourteen generations – strangely through Joseph who one might suppose had nothing to do with it. Leaving aside the fascinating debates on how these ancestral lines might be reconciled, the common thread is the numerology in which the letters DVD sum to fourteen. (By chance, Bach had the same number, and probably embedded it in the B Minor Mass.) The longer genealogy goes back in 3 x 14 = 42 generations to Adam. Six and seven make a perfect multiplication.

The famous six days of the creationists likewise celebrate Seven as another augmented sixth. Adding on one to perfection allows the creator to rest on the Babylonian day of Saturn. But Nature, the awkward customer, never quite fitted the simple story. Second impressions began in that unplanned and unexpected European revolution, when mathematics made its comeback from Asia.

Sorted

Copernicus killed the classical Seven, swapping the roles of the Earth and the Sun and demoting the Moon. When Herschel identified Uranus in 1781, seven planets were temporarily restored. (Uranus had been recorded earlier, and so presumably had been seen, though not noted as a planet, by the naked eye of antiquity.) But it was not to be for long; there were puzzles in the orbit of Uranus which were solved by the observation of Neptune, thus adding an eighth. The next problem was a tiny discrepancy in the orbit of Mercury: some people thought there must likewise be a ninth planet, very close to the sun. But Einstein, in what by this count must be about the seventh impression of the cosmos, was able to explain Mercury's orbit by a wonderful calculation from his new theory of general relativity. This superseded Newton's law of gravity, with the gravitational constant now dictating the curvature of space-time by mass.

The triumphs of unification in physics are often emphasised: Newton for terrestrial and celestial gravity, Faraday
and Maxwell for electricity, magnetism and optics. But perhaps as important is the *abandonment* of attempts at simple
explanations for other, more complex phenomena. For astronomy this meant a slow acceptance that there is no simple law
about which planets must exist. The solar system has essentially random elements in its formation, and so very complicated details governed by deeper and simpler laws of physics.
Kepler had a theory that the planetary distances fitted the
classical platonic solids. It was a beautiful theory, but now seen
not merely as untrue, but as being on quite the wrong level
on which to expect such simplicity. Newton himself probably
did not accept this abandonment, being a last medievalist, a
would-be Magus of the East, fascinated by patterns in historical chronology. A trace of this numerology is suggested by
Newton's inclusion of the Indian 'indigo' in the spectrum to
augment the six to seven colours. Indigo, the blue-jeans dye
that few would clearly distinguish from blue, is no more
distinct a colour than cyan or yellow-green but it makes the
visible spectrum parallel the seven notes from A to G.

There is continuity, not discreteness, in the spectrum of
sunlight, and astrophysics now likewise sees continuity and
chance in the solar system and its collisions: observations of
other stars have now confirmed that it is just one of many
solar systems. The line-up of nearby stars is also a matter of
chance, so that the constellations, despite all the striking
anthropomorphy of Orion, are random effects; the twelve-
ness of signs in the plane of the solar system is a product of

the eye. The patterns of nearby galaxies, clusters of galaxies and superclusters are similar. It is necessary to go further and deeper to find simple structure: back to the uniformity of the Big Bang with its just sufficient primitive fluctuations to explain the present universe's lumpiness.

Going back to the beginning is not as long a journey as some have conceived. At 13.7 billion years, the universe has turned out ridiculously young by Hindu standards, being only about five times older than some of the DNA sequences which we are the latest things to replicate. Life could hardly have started much quicker; it only had to wait for some earlier star to explode to generate the heavy elements. Astronomy is a story of dynamic change, and even now the universe ages, expands and cools by more than 1 part in a billion during a human lifetime. The light from remote galaxies is therefore slightly dimmer than it was in antiquity.

Nevertheless, the Pleiades shine as brightly now as ever the Seven Sisters did, and as the nearest star-cluster in the galaxy, form a vital link in the huge problem of establishing astronomical distances and that 13.7-billion-year figure. Out of centuries of such meticulous and ingenious measurements come *Just Six Numbers*, in the title of Martin Rees's book, which characterise the shape and form of the cosmos, and all springing from its origin. They total a tidy six: but they still present an unsatisfied Seven-ness of untidy loose ends and unsolved questions.

Six is also to be found in the fundamental physics which can be explored in terrestrial experiment. The element uranium was identified in 1789 and named in honour of recently discovered Uranus. Over the next 100 years, it was found to end a roughly periodic table of 92 apparently indestructible elements with various gaps and anomalies, but an integer pattern. Then, in the luminescence of *fin-de-siècle* France, Becquerel and the Curies recalled the banished transmutations of the alchemists. Breaking down the nucleus explained why atomic weights were sometimes not nearly integers (because of isotopes) and even those of pure isotopes not quite integers (because of binding energy and $E = mc^2$). Now the zoo of 92 elements is not expected to be neat and tidy in its mathematical properties.

Instead, after another 100 years, the sprawling pages of chemical properties can at least in principle all be explained as complex effects of six pairs of more primitive beasts, three electron-like pairs and three quark pairs. As it happens, there are also six dimensions for the forces: one for electromagnetic force, two for the weak force (which has a broken two-dimensional symmetry called isospin) and three for the colour force of quantum chromodynamics. This is the *Standard Model*.

But – seventhly – there is the question of the masses and strengths of the particles and forces in the Standard Model. These also involve numbers, and they are not neat integers. Heigh ho! The restless materialist Morlocks have to work even on the seventh day.

Stoney ground

What are these masses and strengths? Ultimately all physical measurements are ratios just as 28 and 365 are based on ratios. Superficially, there are sixes and sevens in the statement that Newton's gravitational constant G is about $6.6742 \times 10^{-11} m^3 kg^{-1} s^{-2}$. But they have no fundamental significance; this is as much a statement about metres and kilograms as about gravity, and these are arbitrary human conventions. (The metre was originally based on the distance from the pole to the equator through Paris.) By changing the units of length, mass and time, G can be made to take any value you like. You may choose units to make it 1, and the speed of light also to be 1. In 1881 the Irish physicist George Stoney realised that another natural choice could be derived from electricity. This was quite a magic guess as the electrons had not then been measured directly, but he was a fine guess-er and it was he who gave the hypothetical electron its name. Using these choices, there are natural units for length, time and mass. Any other physical quantity can be expressed in terms of these units as a dimensionless ratio, a pure number.

Just such a new physical quantity arose in 1899–1900 when Max Planck defined the *quantum*, and thereby his own set of natural physical units. I have earlier called the Planck quantum the fundamental unit of 'existence'. The proper technical term is 'action', but this meaningless word does not convey its radicalism as a four-dimensional idea. It measures a quantity of energy persisting for a quantity of time. H. G. Wells in 1896 had a preview of the necessity of a four-dimensional measure of reality in *The Time Machine*, where the inventor asks: 'Can a cube that does not last for any time at all have a real existence?' Quantum mechanics shows that it is 'action' – not energy, but energy times duration – which comes in integer units. High energy goes with very short time, and so very short wavelengths.

The early years of the twentieth century showed how the Planck quantum related to both light and matter and explained much that was previously inexplicable. Did this unity cohere with Stoney's units? It was not far off, but not identical either. In the early days of quantum physics, the Planck quantum seemed to be $\sqrt{136}$ in Stoney units. Equivalently, the strength of electric charge was $1/\sqrt{136}$ in Planck units.

The mathematical physicist Arthur Eddington was particularly impressed by this integer 136. His underlying idea was that it must be fixed by timeless and logically absolute properties of mathematics, or as Eddington put it poetically in his writing for the public, God is a pure mathematician. After he had explained why it must be 136, the number turned out to be 137. Eddington famously added on one to the divine creation, and explained this augmented 136 instead. More accurate measurement has shown it not to be an integer at all, but about 137.036. This was a magic guess which did not come off.

Other people have since made guesses at a mathematical origin for this number, which is so basic to the properties of all the matter we see and touch, but as yet these have not served to give a theory explaining it. One problem with isolated formulas is that any number whatsoever could be fitted, to a good approximation, by some plausible-looking combination. It is a much denser wood for walking in than we met in fitting fractions to $\sqrt{2}$. The 137 question continues to fascinate, but in a modern context it is only one item in a long menu of seven-ish problems. Foremost amongst these is why the masses are so *small*, as can be seen by expressing them in Planck's units.

The Planck *length* gives the scale of a black hole which fits inside its own quantum wavelength: this is very small at 10^{-33} cm, 10^{-20} of a nuclear diameter. In scale it relates to a nucleus as a nucleus does to a laboratory. That length scale gives a bound below which nothing sensible can be said about space-time: at this level space-time seems to break down into a froth of quantum black holes. It is a signal that some new theory, embracing both gravity and quantum theory, is needed. The Planck *time*, 10^{-43} seconds, is similar. In contrast, the Planck *mass*, which is the mass of such a quantum black hole, is so large that it is on the edge of our awareness. It is the mass of a speck of dust: a flea has a mass of about 4000 Planck units. By $E = mc^2$, the corresponding Planck *energy* is well within the human scale: it is that of a car bomb, about 500kg of TNT.

By the standards of electrons, these are gigantic masses and energies. This observation is better put the other way: the masses of electrons are inexplicably, absurdly small at around 10^{-22} Planck units. Although we usually say gravity is extremely 'weak', it only seems so because the masses of particles are so small. Not all are as small as the electron: the proton has a mass about 1836 times greater, the neutron slightly more, but by Planck's absolute standard these are still very small. In fact, through the success of the quark theory described in Chapter 3, the masses of the proton and neutron are no longer fundamental; the quark masses are now more relevant, and the lowest quark masses are closer to the electron mass. But this does not yet bring the scheme into a recognisable pattern, nor explain the overall small scale: the spectrum of masses still runs from neutrinos (with mass parameters perhaps 10^{-6} that of an electron) to the top quark (of order 10^5 greater than an electron).

One simplification of this picture came in the 1970s from combining electromagnetism and the weak force. But the unified 'electro-weak' force still requires many pieces of information to describe its properties. The neutrinos, like shadows of the electron with no charge and almost no mass, but essential to the weak force, are particularly mysterious and hard to measure.

Another important development allows the masses of all the different particles to be interpreted in terms of their interaction with a single field, that of the 'Higgs boson', with zero spin. This means that the mysterious overall small scale of the masses can be related to the properties of this one boson. But the Higgs boson is not seen directly and is very different from other entities in quantum field theory, so this is much more indirect and theoretical a statement. It gives new predictions and is one area where there may be major advances in the light of new experiments with the powerful Large Hadron Collider at CERN.

At present there are still some 29 numbers, counting ten for unmeasured neutrino masses, which are essentially unexplained. The colour force is perfectly symmetric but the weak force is all over the place. Not all pure mathematicians would agree, on the evidence so far, that this mess is just divine. However, Eddington's view is not the only possible interpretation of the Standard Model: there is another in which it is not pure mathematics at all.

Meaning of life

All atoms, all the physical and chemical properties of matter, depend in a complicated way on the values of these Standard Model parameters. The properties of carbon, gases, liquids, rocks and bones are all functions of them. There are reasonable arguments that, for this reason, life forms could not be very different from those seen on Earth. I once heard a talk by John Barrow explaining this and cunningly making the number 42 appear as the key parameter. As everyone knew from the *Hitchhiker's Guide to the Galaxy*, 42 gives the answer to the meaning of life. Naturally it is: the product of a perfect 6 and an awkward 7, the symbol of science.

Maybe that 42, and so human existence, comes from pure mathematics alone, as Eddington thought. But there is a radically different possible explanation. These numbers might arise as a selection effect, in that conscious beings could never see anything very different. It turns out that certain crucial features of the universe we see are strongly dependent on the values of the Standard Model numbers: having stars at all depends on that electromagnetic constant being very near that $\sqrt{137}$. This point of view was formalised and explored by John Barrow and Frank Tipler in the early

1980s as *The Anthropic Cosmological Principle*.

This principle suggests that the parameters of fundamental physics could have been quite different, but if they had been different there would have been no *anthropos* to see them. We no longer expect the solar system to be based on simple integers, but we do know it must have properties which allow life to exist on one planet, otherwise we should not see it. In the anthropic view, the important thing about 137.036 is that it defines a universe in which conscious beings can evolve to witness it.

The universe, in this view, is like just one of many possible Sudoku puzzles. But in what sense could there be 'other' universes, with other solutions, if the universe (in German, *das All*) is all there is? There are several possible versions. One is that of a creator who chose just the right values out of all those other universes that could have been chosen. Less flatteringly, we could be the by-product of some far more important experiment going on somewhere else; it might be a dry run, or a failed trial. With around a hundred billion galaxies, each with around a hundred billion stars, and with trillions of years yet to go, it is hard to imagine that it was all organised for our benefit. Others imagine game players, rather like the squabbling gods of classical antiquity, experimenting with the universe. The canny humorist Miles Kington, who writes in the *Independent*, supplies regular updates on the committee meetings of the United Deities. Given that the Standard Model looks rather like something designed by a committee, with compromise clauses and apparently purposeless last-minute amendments, Kington's picture might offer a clue to how it came about. But rather than assume a Designer Universe, there are other possibilities. There could be innumerable disconnected bubbles of space-time, of which our universe is just one, within a still vaster universe behind the scenes. The physical parameters would take different, random, values in each bubble. Alternatively, such universes might come into being one after another, or exist in some parallel 'multiverse'.

The anthropic principle does not seem to me a complete answer. Even if the values of the parameters could be explained by such a selection principle, what explanation can be given for the ambient 29-dimensional space? What are the Sudoku *rules* to which this universe is a particular randomly generated, or divinely hand-crafted, or computer-game-played solution? Why, for instance, is there a Six-ness of the integers one, two, three for forces, and three generations of particles? Is that Six-ness some logical necessity? There might be as yet unseen connections between the parameters, and so not 29 parameters but many fewer. Or there might be more para-meters, as yet unseen because they are small or zero. It is a puzzle that the strong force, unlike the weak force, is symmetric in time: is that symmetry a random zero, a selected zero or something that must be zero? Going further, the very word 'random' requires some theory of a probability distribution for what the values could have been, and hence some grander theory behind the universe, analogous to the laws of physics behind the solar system.

In the dark

Recent surprises on the large scale show the danger of premature guessing.

Although the question goes back to the 1930s, it came to a head in the 1990s. The gravitational motion of the galaxies cannot be explained by the masses of their stars and other visible matter. There seems to be much more mass in them than shows up in anything visible. The consistency and detail now achieved through space telescopes and gigantic computer calculations, gives a firm picture of the existence throughout the universe of 'dark matter' — perhaps better called transparent or invisible matter — which is still completely mysterious, as it seems to interact only through gravity. Dark matter had never been predicted as an aspect of the Standard Model. It is a real embarrassment and a warning against scientific hubris: a huge awkward Seventh form of matter to add to the neat six, which makes up as much as 90% of the total mass of the universe.

Analysis of very distant galaxies has also made it possible to measure the accelerating pace of the universe's expansion, explicable only by a further feature of the cosmos known as 'dark energy'. This is a completely different story, going back to Einstein, and a good example of how noughth, first, second, and third impressions of science can succeed each other, for even Einstein had to revise in light of increasing knowledge, and now ideas are being revised back again. Climate change sceptics often pour scorn on the apparent revision from questions of a forthcoming next ice age to an urgent prediction of warming; one can only say that it is the nature of science that new evidence may completely supersede the old.

In 1917, Einstein put forward a picture of the cosmos consistent with his 1915 equations for general relativity. But to fit a universe whose shape did not change with time, as then seemed sensible, he added an extra term to the equations. At that time nothing was known of the galaxies beyond our Milky Way, but in the 1920s astronomy rapidly revealed them and the overall expansion of the universe. Einstein's original equations now fitted these discoveries much better and the motivation for the added 'cosmological' term was lost. Einstein is said to have described it as his greatest mistake. Nevertheless, that extra term had a good rationale: it was the one modification consistent with the principle of depending only on the absolute properties of the geometry. This extra term says that space-time possesses an intrinsic curvature: nothing to do with matter, just pure geometry. From the 1920s to the 1990s it was taken into account and regarded as a theoretically possible, but probably not significant extension of general relativity. The amazing fine detail revealed by space-telescope observations since the late 1990s points clearly to exactly this term being needed after all. The curvature involved is extremely slight, of the order of ten billion light years, but the effect is cumulative and has come to dominate the geometry of the universe at its present age.

For some reason it is thought very important not to prejudge the issue by identifying the astronomical observations as a clear confirmation of the effect Einstein allowed for. The non-commital term 'dark energy' is used, and much effort is made in the analysis to keep all options open. But it is to my mind misleading: the observed curvature effect is quite different, requiring a gigantic 'negative pressure', from anything produced by the energy of matter.

So it appears that Planck's length of 10^{-33}cm, identified as fundamental for over 100 years, has recently been joined by an equally fundamental 10^{28}cm cosmological radius. There is a new duality of beginning and end, large and small, emerging from this music of the spheres. From bass to treble this range covers some 200 octaves of frequency, with a tune more like Stockhausen than Mozart, and the particles tinkling in about 40 still inexplicable octaves in the upper middle register. As it happens, the expression 'sphere' is apposite. The most fundamental Planck unit seems to be that of the spherical *area* of a Planck-mass black hole. Curvature is also naturally expressed in terms of area, so Einstein's cosmological constant is also most naturally thus expressed. The ratio of these two areas gives a pure number of the order of 10^{123}. At present, this gigantic number seems to be the most fundamental fact of physics, more basic than the famous 137, characterising the very nature of space-time even before particles come into it, but completely unexplained: a super-sized Seven to be sorted.

There is a further surprise: the combined effect of known matter, dark matter, and the intrinsic curvature makes the universe *spatially flat* on a cosmic scale. This might appear to contradict the idea of curved space-time. It does not: the four-dimensional space-time is indeed curved, and the universe everywhere is bumpy and wavy and holey with its stars and galaxies. But large-scale slices through the universe, obtained by looking at sections with the same age and temperature, seem to form an overall flat three-dimensional space. In particular, it is not the 3-sphere that would be just as consistent with Einstein's equations. Martin Rees's *Just Six Numbers* emphasises that such flatness is needed to get a long enough time for galaxies and stars to form but not to collapse. This zero spatial curvature, one of his 'six numbers', is an unexplained zero, and seems to be a knife-edge coincidence. The mechanism of cosmic *inflation* is supposed to explain this flatness as something fixed at the Planck-time age of the universe, but this cannot yet be called a complete theory.

Maybe there are surprises to be found also at the small scale.

In the quark?

The awkward Seven-ness of particles and forces suggests that there may be a deeper layer to be found. Just as the 92 elements have yielded to a simpler Standard Model of electrons and quarks, it is possible that these in turn may be reduced to a yet more fundamental level, from which the puzzling numbers can be explained. Such deeper structure is relevant whether or not an 'anthropic' answer is expected.

It's like going to sea in a sieve, buoyed up by the success of earlier inspired guesses, but as yet without a basis in experiment. Stoney's electron, Planck's constant, Bohr's atomic model, general relativity, quarks and much else started life on the basis of simplicity of explanation rather than on massive databases. String theory dominates this latest voyage of discovery.

The effect on mathematics is striking; to read Constance Reid's survey of number theory in 1956 is to see how it lacked the invigoration and challenge that has come from this ever-expanding programme of research. Physicists' voracious appetite for advanced mathematical structure has brought together aspects of new geometry and algebra that might too easily have remained isolated in tunnel vision. It has restored a kind of creative magic and adventure to mathematical vision.

Some express impatience with the fact that despite 30 years of this effervescence, no definitive explanatory theory has emerged. The prominent physicist Lee Smolin, for instance, has said that string theory should be given up. But his own suggestion is still something similar, being based on the idea which takes off from the success of general relativity: to get everything into geometry. He favours a theory of *braids* to explain the Six-ness of the quarks and electrons: these are stringy knots – knotted surfaces rather than knotted circles. It is hard not to make the comparison with Kelvin's elegant (but completely wrong) theory of chemical elements as discretely different knots, before the truth was found in quantum theory. There are other 'preon' theories which are not based on strings. Hong-Mo Chan and Sheung-Tsun Tsou have a 'Model behind the Standard Model' with the idea of breaking electrons and quarks down into a deeper level, based on their Six-ness. This scheme, which reduces the number of parameters, is commended by Roger Penrose in his book *The Road to Reality*, in the course of criticism of the assumption that string theory in its present form must hold the answer.

The power of mathematical synthesis is such that what now look like either/or contradictions or doublethink may fade when seen in a new light. The radically different twistor description, as mentioned in Chapter 4, offers the possibility of such a new way of looking at the Standard Model. The Higgs boson description of mass has the effect of defining a theory which is in a sense a more fundamental level of the Standard Model, and which is, like light, scale-independent. The usual space-time description does not express this symmetry efficiently; nor does it express the asymmetry of the weak force well. Twistor space could be looked at more seriously as the right framework for this more fundamental level.

Often it is wrongly assumed that only light-like entities can be expressed in terms of twistors. This is not true: the point is that the breaking of conformal invariance is made explicit by twistor geometry. It is seen in the 'line at infinity'– something finitely described by twistor space, very like the horizon in perspective drawing. The concept of a constant mass or length, as well as the origin and final shape of the universe, must go into that horizon. Witten's use of twistors in string theory has only used the simplest connection of twistors with Minkowski space, not using their full potential. The same is true, so far, of the twistor-based system described in Chapter 4 for working out simple colour-force processes. But if there is to be serious use of twistor space, then full attention must be given to this structure 'at infinity', which has completely new features, different from anything in space-time. Physicists are so far understandably reluctant to adopt such an unfamiliar perspective, and place more trust in the assumed reality of space-time points. Perhaps a turn of thought greater than that of 1989, greater than that of the fourth century, would be needed.

The scientific record of the past century suggests that this chapter will soon look like faded pages from Eddington, and that another hundred years could bring a synthesis currently unimaginable. A more pessimistic reading of the historical record suggests a desertified, flooded, depleted, cluster-bombed, radioactive planet for 2107. Looking on the bright side, there are plenty more galaxies and lots of time, so the All is not lost.

The final reckoning

'Seventhly' is demanding enough on the reader – and 'eighthly' would be unbearable. I will mention only that *eight-dimensional supersymmetry* gives an approach to quantum gravity which seems to hold some vital clue, though still on the magic guessing level. So maybe Eight will be the number of a final unification of the Four-ness of gravity with the Two-ness of quantum mechanics. But we will turn to another and simpler Eight, again by going back to the wonders of antiquity. The sifting of the world's intractable sand began then – quite literally, as Archimedes started to measure the universe with its grains. His *Sand-Reckoner* initiated the idea that has now evolved into 10^{123} – sorting out the ratio of very large to very small.

The words *Ten thousand times ten thousand, in sparkling raiment bright* suggest an unusual influence in Christian hymnody, anomalous as the pagan gods obstinately resisting eviction from the days of the week. This is because that figure is not three-based but eight-based. It is the *myriad of myriads* which Archimedes used as the basis of his description of large numbers – far larger than the small-town vision of 144,000 in the Christian revelation. Archimedes was able to define numbers up to the myriad-myriadth power of a myriad myriads, in modern notation $(10^8)^{10^8} = 10^{8 \times 10^8}$ – a one followed by 800 million noughts. For the volume of the universe, Archimedes then derived a figure. The space extending out to the stars could be reckoned as equal to 10^{57} grains of sand.

Astronomical numbers are generally within the range of a *googol*, this being defined as 10^{100}. The googol has never actually been used in any serious context. It is famous for being famous, being quoted and requoted from one popular book to another, especially after its homonym became the business of billionaires. But it gives a reasonable picture of the scale of the material world. Such is the present reckoning, an updating of Archimedes. And the idea of reckoning leads to a new angle on numbers and the physical world. The gateway, which has also made a trillion-dollar business, is the *computer*, based on –

8

Sound Bytes

Computers work with zeroes and ones. Digital art installations regularly show the inhuman reductionist nature of science, etc. etc., with images rendered in chilly zeroes and frigid ones. Sadie Plant wrote on *Zeroes and Ones: Digital Women and the New Technoculture*. Back in the 1950s, zeroes and ones were still cutting-edge news and Constance Reid explained them in Chapter 2 of *From Zero to Infinity*. Now everyone knows that binary digits, or *bits* for short, underpin electronic computing. And yet:—

Achtung! Achtung!

As Constance Reid pointed out, you might also say that computers use numbers in base 4, 8, or 16. Use of these bases only amounts to employing slightly different *names* for the binary numbers. Take the number we usually write as 2885, which is 2048 + 512 + 256 + 64 + 4 + 1 and so represented in binary as 101101000101.

> Writing it as 10 11 01 00 01 01 is completely equivalent to using base-4 representation, with 00, 01, 10, 11 as numerals instead of 0, 1, 2, 3.
>
> Writing it as 101 101 000 101 is completely equivalent to using base-8 representation, using 000, 001, 010... 111 as numerals instead of the usual octal 0, 1, 2 ... 7.
>
> Writing it as 1011 0100 0101 is completely equivalent to using the base-16 representation, using 0000, 0001, 0010... 1110, 1111 instead of the hexadecimal 0, 1, 2, 3... E, F.

The same argument applies to any other power of 2. Some early computers used teleprinters for input and output, and so used 5-bit sequences for alphanumerical symbols; they were effectively using base 32. Bases of 8 and 16 became more popular, and base 16 became standard for microprocessors. But the number Eight has made a comeback, not as the octal representation but through the fact that $256 = 2^8$. This means that a sequence of *eight* binary digits is effectively a numeral in base 256. By the 1970s it was standard to work in such 8-bit units, called *bytes*. A byte can be written as a pair of hexadecimal symbols.

Another aspect of computing, increasingly important with the global development of the Internet, is that the coding of alphanumeric symbols has evolved from those original

5-bit teleprinter symbols into ASCII (American Standard Code for Information Interchange) and now includes all world writing systems, including Chinese ideograms, in the Unicode system. This expansion also uses a base of 256. Meanwhile the internal processing chips themselves have come to use 64-bit registers, consisting of eight bytes each of eight bits.

So Eight is the number of *computing*. It was prefigured by Archimedes' organisation by myriad-myriads, and also by the 8 x 8 chessboard for well-calculated strategy. Readers of my earlier book *Alan Turing: the Enigma* will know the drama of the Eighth Square, borrowed from Lewis Carroll.

The hugeness of possibilities within a computer is even greater than the number of possible bridge hands. Just one 64-bit register, a microscopic component of a computer, can hold any one of 2^{64} different numbers. At a thousand trillion operations a second (the fastest in 2006) it would take five hours to work through them all. There are 2^{128} possible states of *two* such registers, and even using ten billion chips in parallel, it would take a million years to work through them. Therefore most simple additions of two 64-bit registers have never been performed.

That super-eight number 2^{64} was used at least twice in pre-computer days to illustrate the enormous scale of combinatorial numbers arising even from simple pictures. A legend tells of the inventor of chess, who asked for a reward consisting of a grain of rice, redoubled for each square of the chessboard, a request to which the Indian king rashly agreed. A nineteenth-century French mathematician, Edouard

Lucas, devised a puzzle called the Tower of Hanoi, looking innocuous enough, but soluble only in $2^{64} - 1$ moves. But the formidable power of modern computing comes not from the zeroes and ones of binary numbers, nor even from the huge number of ways of combining them.

> Fact: You need to understand binary numbers to know how computers work.
> Deeper fact: You don't need to understand binary numbers to know how computers work.

The power of computers comes from the potential to run *programs*.

Under instruction

Although Eight has thus played a practical part in computing, something much deeper comes from thinking about the way that eight arises as the cube of two. In Chapter 4, we saw cubing as a *function*, pictured as a *graph*. This holds the idea of cubing as a platonic relationship: the function is defined to *exist*. The cube and its inverse, the cube root, are both simply *there*. But there is quite another way of thinking of cubing, in which the symmetry of doing and undoing is completely broken. In writing $2^3 = 8$, that 3 can be thought of as an *instruction*: start with 1 and multiply by 2 three times.

Think of this as a cookery recipe. It is no more difficult to write a recipe for 2^n, where n is any natural number. But doubling three times comes close to the limit of what you can do without keeping a careful note of how many doublings you have done. So a wise cook will do something like this to keep a count of the doublings:

POWER BREAKFAST RECIPE

Step 1: Put 1 into the pot, put n in the pan.

Step 2: Double the number in the pot, decrease the number in the pan by 1.

Step 3: If the number in the pan is not 0, go back to step 2; otherwise take the number out of the pot.

The number that emerges is just 2 to the power of n. Now there is no problem if the phone rings in the middle. This flexibility imposes a price: the cook must apply a test and go to a step which depends on the outcome. But this is natural for cookery: it embodies the idea of 'bring to the boil'. It is simple for numbers as well.

It is no harder to write a recipe for m^n, by changing 'doubling' to 'multiplying by m'. So the operation of working out a power, or *exponentiation*, can be reduced to a sequence of easier *multiplications*, augmented by the business of testing and jumping. Is this the end of the story? No, because multiplication is not really a single step. It also needs a recipe to reduce it to a sequence of even easier *additions*. In turn addition can also be reduced to the repeating of the very basic operation of *incrementing by one*. Putting these recipes together, the business of working out m^n can be reduced to a succession of primitive add-on-one operations, but with constant testing and jumping required.

To put these recipes together, another principle is needed. It is that of 'here's one I made yesterday', as tirelessly used in cookery demonstrations. Imagine that the recipe for addition has been written out and is available on demand. Then the recipe for multiplication goes ahead by instructing the cook at various points to 'look up the recipe for addition, and follow it'. The recipe for exponentiation similarly contains instructions for getting and using the multiplication recipe.

This is, basically, just how computer programs break everything down into very simple operations, each to be implemented electronically in a trillionth of a second. Yet it was conceived before computers existed.

We have already seen something of the background to this in Chapter 1. The meticulous logic, such as Russell pursued, set the stage. It put mathematicians into the mind-set of deriving conclusions from axioms *mechanically*, completely without imagination. Indeed Hardy actually used the word 'machine', rather pejoratively, to describe this attempt to make mathematics into a complete set of formal rules. Gödel's work in 1931 had introduced the idea of *coding*, and mathematics thereafter was edging towards the kind of logic needed for computers, the logic of actually performing operations, not just asserting truths.

Yet this recipe-book tradition had always been present in mathematics. An older, mathematical word for method is 'algorithm'. The word is nothing to do with Al Gore, but comes from al-Khwarizmi in medieval Iran, long before can-do America was thought of. He in turn drew on the Greek mathematicians Archimedes and Diophantus who expressed their discoveries as methods for solving problems. Some of

Euclid's statements were also put in this way. Fermat's famous 'last theorem' conjecture, written in the margin of his copy of Diophantus, is usually given in the form of a 'non-existence' statement, but actually what he wrote was that 'it is impossible to separate a cube into two cubes...' using the Diophantine language of performing an operation. Galois and Abel drew on this tradition in showing that the methods for solving quadratic, cubic and quartic equations could not be extended to quintic equations. In the twentieth century, the 'constructivist' philosophy, as a development of intuitionism, had already challenged the Platonic ideal of pure 'existence' and was all part of the shake-up of logic.

Nevertheless, it needed someone quite new to crystallise this. If you are a believer in genetic connections, you will find it significant that George Stoney's second cousin's granddaughter Sara Stoney had a son who turned out to be just this person. Alan Turing, when just turning 23, had the idea soon called *Turing machines* – essentially like the recipes described above, reducing everything to the simplest possible components, but showing a potential for encompassing enormous tasks.

Turing's work was published in 1936. It was stimulated by what Gödel had done in 1931. It had nothing to do with IBM, and was not done for profit. Nor was it motivated by current needs for computing in science. The early 1930s marked a point when enormous calculations for quantum chemistry were put on the agenda. Turing certainly knew of such problems, but his new cuisine was not devised to solve them. An interesting question is whether Turing was guided by the

Marxist Zeitgeist which emerged in Lancelot Hogben's writing at just this period. I think not: certainly he had a hands-on approach to mathematics, but this was a distinctively individualistic passion rather than something drawn from an ideology about workers by hand and brain. Once that passion had emerged in the form of Turing machines, it inspired others who were alive to its potential.

Really really large numbers

In 1938, as war approached, two mathematicians, Janos von Neumann from Hungary and Stanislaw Ulam from Poland, were crossing the Atlantic on the *Georgic* liner. Ulam wrote later that von Neumann 'proposed a game to me... writing down on a piece of paper as big a number as we could, defining it by a method which indeed has something to do with some schemata of Turing's... Von Neumann mentioned to me Turing's name several times in 1939 in conversations concerning mechanical ways to develop formal mathematical systems.'

To write down big numbers on a piece of paper, using Turing machines, they very likely started with the following idea. It illustrates the power of programs by getting some *really really large* numbers from just a few lines. We simply go in the opposite direction, and treat the exponentiation recipe as the 'one I made yesterday', to be read and followed as part of a much more ambitious recipe:

TOWER OF STRENGTH RECIPE

Step 1: Put 1 in the pot, put n in the pan.

Step 2: Get the Power Breakfast recipe. Use it to replace the number in the pot by 2 to the power of the number in the pot. Then subtract 1 from the pan.

Step 3: If the number in the pan is not 0, go back to step 2; otherwise take the number out of the pot.

What kind of numbers come out of the pot? They rapidly become so large that new notation is necessary even to describe them. First, we introduce a single arrow ↑ for exponentiation, writing $m \uparrow n$ instead of m^n. Next, we define the *double* arrow ↑↑ by the rules: $2 \uparrow\uparrow 0 = 1$, and then $2 \uparrow\uparrow n = 2 \uparrow (2 \uparrow\uparrow (n-1))$. Then following this rule:

$$2 \uparrow\uparrow 1 = 2^1 = 2$$
$$2 \uparrow\uparrow 2 = 2^2 = 4$$

This gives a new way in which two and two make four!

$$2 \uparrow\uparrow 3 = 2^4 = 16$$

Now the repeated exponentiation begins to get serious:

$$2 \uparrow\uparrow 4 = 2^{16} = 65536$$
$$2 \uparrow\uparrow 5 = 2^{65536}$$

After this we cannot even write out the exponent and have to leave it as:

$$2 \uparrow\uparrow 6 = 2^{2^{65536}}$$

Without the use of the double arrow notation, any further numbers have to be written as *towers of powers*. $2 \uparrow\uparrow n$ is a tower of twos of height n, like this:

But even this needs a forest of brackets to make it clear that the evaluation must start at the top and work downwards. This is why the arrow notation is valuable.

The number 65536 can be visualised: a town, a big crowd, or the number of bits in a small image. The next number, 2 ↑↑ 5, is about 20000 digits long in decimal notation, far greater than astronomical numbers. It is comparable with the number of *possible* images that the eye and brain must be able to cope with. 2 ↑↑ 5 is not as great as Archimedes' largest number, but 2 ↑↑ 6 leaps ahead, far beyond any mental picture.

DEADLY: Arrange in order of size the following numbers:

(A) Archimedes' largest number,

(B) Eddington's guess for the number of particles in the universe, $2 \times 136 \times 2^{256}$,

(C) a googol,

(D) a googolplex, defined as 10 to the power of a googol,

(E) the largest number you can make with four fours,

(F) Penrose's number measuring the specialness of the Big Bang, $10^{10^{123}}$,

(G) 2 ↑↑ 5, (H) 2 ↑↑ 6.

But this is only the beginning of what can be done with the arrows, or equivalently, what can be done with a few lines of programming. Taking this just one stage further, you can go on to write a simple recipe for *really really really* large numbers. Its outputs cannot be expressed with conventional mathematical notation, even allowing for towers of powers. The recipe is equivalent to the rules: $2 \uparrow\uparrow\uparrow 0 = 1$ and $2 \uparrow\uparrow\uparrow n = 2 \uparrow\uparrow (2 \uparrow\uparrow\uparrow (n - 1))$.

Then $2 \uparrow\uparrow\uparrow 1 = 2$ and $2 \uparrow\uparrow\uparrow 2 = 4$ again. Next,

$$2 \uparrow\uparrow\uparrow 3 = 2 \uparrow\uparrow 4 = 65536,$$

$2 \uparrow\uparrow\uparrow 4$ is a tower of twos 16 high,

$2 \uparrow\uparrow\uparrow 5$ is a tower of twos 65536 high,

and after that you can't even imagine the size of the tower of twos.

The arrow notation can be extended without limit. $2 \uparrow\uparrow \ldots \uparrow\uparrow 2$ is always 4. $2 \uparrow\uparrow\uparrow\uparrow 3$ is $2 \uparrow\uparrow\uparrow 4$, on the edge of comprehensibility, but $2 \uparrow\uparrow\uparrow\uparrow 4 = 2 \uparrow\uparrow\uparrow\uparrow\uparrow 3$ and all further numbers of this kind defy imagination.

The universal machine

If he had been sharing that voyage in 1938, and had overheard this pair of *mitteleuropäische* professors excitedly scribbling Turing machines in their deckchairs, Lancelot Hogben might have considered their competition a frivolous exercise in abstract mathematics. Yet the principle of programs calling other programs has turned out useful for even more millions of people than have ever read Hogben's book, and has changed the very nature of work by hand and brain. That is because it is the principle on which modern computers work. Underlying it is the fundamental discovery that Turing made in 1936, the concept of a *universal machine*. The modern computer (more strictly defined as a computer with 'modifiable internally stored program') is an embodiment of that concept.

It is easiest to appreciate it with the advantage of hindsight. A computer stores everything in zeroes and ones. The numbers it works with are all in zeroes and ones – but so are the programs. If one program needs to call on another, it needs to fetch and read a sequence of zeroes and ones. This is no different from fetching a number. The technology for storing and communicating the programs can be exactly the same as that for numbers. The cookery demonstration imagined numbers in pots and pans, and recipes in a book, but you could keep the recipes in pots as well. In a word, computers are not kosher. For modern PC users all this is familiar. A click on the mouse may run a program, and so use it as a recipe, but you may just as well download, install, store, copy or compile the program, thus treating it as a file of 0-and-1 data.

In 1936, the idea of encoding operations *on* numbers *as* numbers, which Turing announced, was revolutionary. But it was strongly connected with Gödel's 1931 work, which had encoded statements *about* numbers, *by* numbers. This in turn derived from the logical problems of Russell's sets which are members of themselves. So the power of modern computing does in fact derive from the most rarified of intellectual efforts to find consistency and unity in mathematics.

The upshot of Turing's work was that however complicated the task, it could be reduced to simple operations like adding-on-one, and performed by a single machine. The complexity of the task would go entirely into the logic of the instructions. This is the concept of the modern computer, which, as we would say now, only needs new software in order to switch from one task to a new and completely different one.

Often the word 'digital' is used as if it expressed the one and only essential principle of computer technology. It is indeed true that the digitalisation of information is necessary for it. But that is not sufficient. A universal machine, reading programs and carrying them out, needs a certain level of organisational complexity, and this took over two thousand years of logic to develop. Charles Babbage, often claimed to be the inventor of the computer, developed some first ideas about programming in the 1840s. Ada Lovelace gave a famous account of his most advanced application – the calculation of Bernoulli numbers, as mentioned in Chapter 6. But her explanation shows that their ideas about how to take output and feed it back as input were vague and confused. They never had the central idea that instructions and numbers can be stored and manipulated alike. I would compare the Analytical Engine plans with the tiles of Darb-i Imam: they are amazing and seem to jump out of their time-frame but cannot be said to express the content of modern understanding.

The aristocratic Ada Lovelace, Byron's daughter, gives a particularly vivid example of a woman who seems out of her time-frame. But it is not clear how much she actually contributed to the Analytical Engine. Other women, such as her teacher Mary Somerville, and the later Sofia Kovalevskaya and Emmy Noether, are less celebrated but arguably greater mathematical pioneers.

Turing, in contrast, saw the whole thing. It is not clear whether in 1936 he surveyed the possibilities of constructing a universal machine. The price of universality is that a great many logical operations have to be performed to follow the instructions, and this is not worthwhile unless millions of them can be performed in a second. The speed of electronic components, necessary for this, was not available in 1936. But once Turing learnt electronics from the new technology developed for code-breaking at Bletchley Park, he realised it could be applied to his concept of the universal machine. After emerging from the war against Nazi Germany, which he probably did as much as any individual to win, the first thing Alan Turing did was to draw up a plan.

He was beaten to it. Von Neumann's first design of an electronic computer appeared some months before Turing's own detailed scheme, and von Neumann has largely gained the credit for initiating the post-war computer revolution. (His first program bore the iconic date 8 May 1945.) Von Neumann was a more organised and powerful proponent, but Turing had the sharper and deeper picture of universality. This still took a long time to sink in after 1945, even though the power of von Neumann's and Turing's designs rapidly swept the field. Martin Davis, one of the leading logicians of the past half-century, has written a history of Turing's concept as *The Universal Computer*. He quotes Howard Aiken, the computing chief of the of US Navy, and the person who did most to bring Babbage's ideas to fruition. As late as 1956

Aiken wrote that 'If it should turn out that the basic logic of a machine designed for the numerical solution of differential equations coincide with the logics of a machine intended to make bills for a department store, I would regard this as the most amazing coincidence that I have ever encountered.' This amazing coincidence – a very good statement of the power of the universal machine – is now completely taken for granted.

At first, department stores – and capitalist enterprises generally – were little interested, leaving government and advanced scientific research to realise its potential. As with the Eurabian zero, realisation is just the right word. The internally stored program has conquered though being realised in electronics, and now there is no going back. With enormous advances in scale and speed and reliability, a mass market has been created such as no one dreamt of in early days. Programs for gaming and webcamming, linking a global network of teenage bedrooms, are not exactly what progressive thinkers of the 1930s planned for future society, nor what technocrats of the 1940s and 1950s envisaged as the economic role of computing. Indeed, they go far beyond what Bill Gates saw in 1993. But they follow from the power of the universal machine, and its internally stored programs all calling each other.

Hardware is ephemeral, as Turing saw. Today's personal computer may disappear, and the word 'computer' itself may be overtaken by new developments. Already, mobile phones are (rightly) advertised as showing where computers are going. They may become too small to see, they may be implanted in brains, but universal machines will retain their logical power.

User-friendliness

The huge success of the personal computer market has not been because of the public wising up to binary numbers. Computer users are quite right not to know or care about them. As Turing also explained very clearly in 1947, any conversion between decimal and binary numbers could be done by the machine itself, simply by writing appropriate software. This is just another aspect of universality, and it makes the computer essentially user-friendly.

This is why you don't need to know about binary numbers to use a computer. You do not need to cater to its way of working: it can be instructed to cater to yours. As long as you can tell it exactly what you want, the computer can do it. You could call this the 'Hot Pink' principle. In Chapter 3 I explained how that name is given to the mixture of red, green and blue expressed by the hexadecimal symbols FF69B4, which web-browsers read. But if you want to create a webpage with Hot Pink background, there is no need to remember or look up those symbols yourself. The computer itself can translate the more user-friendly words 'Hot Pink' into hexadecimal form. As Turing put it in 1946, 'The process of constructing instruction tables should be very fascinating. There need be no real danger of it ever becoming a drudge, for any processes that are quite mechanical may be turned over to the machine itself.' From this perception flows the idea of computer languages, which Turing himself started with instructions written in what he called a 'Popular Form'.

Turing wrote in 1946 that 'Every known process has got to be translated into instruction tables.' Turing was far ahead of von Neumann in seeing the importance of software, and this is indeed how it has gone, though with applications that would have surprised the 1940s pioneers. Turing also saw that computer experts would resist user-friendliness and use 'well-chosen gibberish' to explain why things could not be done. Maddening error messages, incompatible files, crashes and connection failures bear out his words. 'Computer says no' is a running joke in the BBC's *Little Britain*. But these are not faults of the underlying concept. If anything, the problem lies in the other direction: the computer can't say no to anyone, in any language. The universal machine is a shameless slut worthy of *Little Britain*, just as friendly to abusers as to users. If instructed, it sends out a million spam emails with spoofing and phishing, or injects Trojan-horse viruses into millions of other computers. It can create firewalls or hack into firewalls, censor the Web or defeat the censorship. It follows instructions.

Only obeying orders

One of the climate-change-sceptic arguments is that the predicted increase in global temperature is 'only a computer model'. This is not a strong argument: successful landings of spacecraft on Mars, and the prediction of protein shapes from DNA sequences, are the results of computer models. The fact that a prediction is run on a computer is not of primary significance. The validity of the model comes from the correctness of its mathematics and physics. Making predictions is the business of science, and computers simply extend the scope of human mental faculties for doing that business. To reject computer models *per se* is to reject science. On the other hand, there are serious questions about how the continuous world of physics can be modelled on the discrete states of computers, and for climate questions – as for the proton model – much hangs on the fineness of the grid, or lattice, of space-time points. Treatment of this question needs its own sophisticated mathematical theory, that of *numerical analysis*.

Conversely, if a theory is wrong then running it on a computer won't make it come right. Astrology does not become scientific by being computer-based. In view of the fact that many people take leave of their senses when computers are involved, there is a good place for scepticism about computer outputs unsupported by reason. The tendency of people to believe in machines is not new: the wartime German authorities could not believe the Enigma was at fault, and preferred to distrust their own U-boat crews. With gigantic software contracts, identity register systems, medical records, on-line voting – computers are quite capable of generating a stream of crime and folly to enhance all the others.

Computers do algorithms. Whether the results are true or false, elegant or ugly, fraudulent or frightening, tedious or puerile, illegal or totally ridiculous, they carry them out. But specifying an algorithm means giving a completely definite, finitely expressed set of instructions. For many tasks it is doubtful whether complete rules can ever be given. An example which causes frequent hilarity in the press is the application of software to filter messages for rude words. But even if we stick to the safety of pure numbers, we encounter tasks which cannot be performed by algorithms and so cannot be done by computers.

When computers can't decide

Just as the idea of the universal machine predated electronic machines, so did the discovery of the *uncomputable*. It was there right at the beginning in Turing's 1936 work, and it had roots in the logical antecedents: Russell's sets which are members of themselves, and Gödel's mathematical statements which refer to themselves. Turing had followed these by considering programs operating on themselves. This led to the very positive idea of the universal machine. But it also established the negative fact that some problems are, in a strict sense, unsolvable.

A program will involve much testing and jumping back and forth between instructions, and it is hard to tell by looking at it that it will not go into a loop of operations from which it will never escape. Define 'debugging' a program to mean checking it to ensure that it will actually terminate, rather than go into such an unending loop – the effect seen with faulty software as a computer 'crash'. Can debugging of programs itself be done by a program? It turns out that this is impossible. Turing's principal result in 1936 can be thought of as showing the non-existence of any completely general program for debugging. It can also be seen as showing that there is no general algorithm for settling mathematical questions.

Why can the apparently technical business of 'debugging' be equated to a problem of pure mathematics? The underlying idea can be seen from our simple cookery recipes. These used a single pot, or register, for a number acting as a counter ticking down to zero. The recipe relied on the fact that we know the properties of numbers, and therefore know that however large the starting value n, it will, if decreased each time, eventually reach zero and so stop the computation. But a slightly more complicated program might have not one but several counters ticking away. At each stage they could be tested not for whether they have reached zero, but for some more complicated mathematical property. Four counters a, b, c, n could be tested for the truth of $a^n + b^n = c^n$. To know whether a program depending on this test would go into an unending loop, you would have to know the truth of Fermat's conjecture. As it happens we do now know this truth, but there are infinitely more statements whose truth is unknown. Having a complete debugging program would be equivalent to being able to decide all of them.

But Turing's theory made it easy to see that no such debugging program could exist. If it did, it could be applied to *itself*, and would produce a self-contradictory tangle. Putting these ideas together, Turing showed that mathematical methods cannot be exhausted by a finite instruction book: creativity will always be required.

Twenty years later, the theory of computation was

re-entering the serious heartland of Number by extending Turing's discovery. Although Constance Reid's 1956 book gave a rather low-level picture of computers doing binary arithmetic, she actually knew better than that. She was connected with another American mathematician, Julia Robinson, a pioneer in this high-level appreciation of computing. Julia Robinson, like Martin Davis, did important work in a long series of discoveries which culminated in 1970 when a young Russian, Yuri Matyesevich, finally showed that there is no algorithm for solving all Diophantine equations. In this way, computing got back to Diophantus, the proponent of the constructive method.

The proof required extraordinary new algebra, in which Fibonacci numbers played a vital part. It led to the discovery of hitherto unsuspected aspects of the theory of numbers. A *polynomial* is a formula using only addition and multiplication: as a result of this work it was found that there is a polynomial in 26 integer variables whose positive values are precisely the prime numbers. There are analogous polynomials for any series of numbers that can be produced by a computer program. As a more vivid picture of this line of research, the Penrose tilings arose from studying an *uncomputable* problem. This is yet another example of how a proof that something can't be done leads to completely new ideas.

Under construction

Practical work with computers has also shed a little new light on those powers and primes of the classical mathematicians.

The world's first universal machine was working in Manchester in 1948. It only had a store of 1024 bits. But this was enough to launch a search for *Mersenne primes*, named after Martin Mersenne, another seventeenth-century figure from Fermat's France. Mersenne primes are those which are just one less than a power of two, and so appear as 111111 . . . 1111 in binary notation; they have particularly interesting properties. One is that they are connected with perfect numbers. A perfect number is equal to the sum of its divisors. The first example is $6 = 1 + 2 + 3$ and the next is $28 = 1 + 2 + 4 + 7 + 14$.

DIFFICULT: The number expressed in binary notation as 1111 . . . 111000. . . 000, with n 1's and $(n - 1)$ 0's, is perfect if 1111. . . 111 is prime.

Only numbers of form $2^p - 1$, where p is prime, can be Mersenne primes. The first of these occur for $p = 2, 3, 5, 7, 13, 17, 19, 31$, as Euler found. These give the primes 3, 7, 31, 127, 8191, 131071, 524387, 2147483647.

DIFFICULT: Constance Reid left this as a 'challenge' to her readers: If n is not prime, then $2^n - 1$ is not prime. (Hint: Express $2^n - 1$ in binary notation.)

Beyond $p = 31$, the question becomes decidedly TRICKY. It is solved by using a special test for the primality of numbers of this particular form. This test was found by the same Edouard Lucas who devised the Tower of Hanoi puzzle. It involves an extension of Fibonacci numbers. Using it, Lucas showed $2^{127} - 1$ to be prime, and this number remained the largest known prime for many decades, until the record fell to the computer. The Lucas test, a little refined, is particularly suitable for the zeroes and ones of computers, and this is why it could be run as virtually the first program on that tiny Manchester prototype. Even though Turing was involved, it didn't find any more, but in 1952 a new prime, $2^{521} - 1$, was found at Cambridge, and a few more followed. The new results were impressive enough to inspire Chapter 6 of Constance Reid's *From Zero to Infinity*.

The subsequent story of Mersenne numbers marks the complete change from the days when the world's few computers were reserved for a tiny élite. Personal computers far more powerful than any computer of the 1950s, and linked through the Internet, allow anyone to join in. In September 2006, the 44th known Mersenne prime was found and $2^{32582657} - 1$ became the largest known prime. The organisation of 'distributed computing' for amateur participation, like the SETI search for messages from outer space, `fightaidsathome.net` and `climateprediction.net`, shows the power of universal user-friendliness. But whether there are infinitely many Mersenne primes, and numerous other such questions, cannot be resolved by computing, however vast the scale.

Those really really large numbers give a sense of how strong a statement it is to say that something is true for all numbers. Never, never, never, even for those unimaginably large numbers defined by the repeated $\uparrow \uparrow \uparrow$, can it be true that $a^n + b^n = c^n$ for n > 2. That truth, proved at last by

Andrew Wiles in 1995, could not have been established by computers. Computers could have *disproved* Fermat if his conjecture had been wrong. They have disposed of less lucky conjectures. Euler, for instance, guessed a generalisation, that just as there is no cube which is the sum of two cubes, there is no fourth power that is the sum of three fourth powers, no fifth power that is the sum of four fifth powers. But he was wrong, because computer search has revealed that $144^5 = 27^5 + 84^5 + 110^5 + 133^5$ and $422481^4 = 95800^4 + 217519^4 + 414560^4$.

The Riemann hypothesis, which is about the behaviour of all prime numbers, however large, is another statement that cannot be shown to be true by computing. But if untrue, it could be shown to be so by numerical calculation. In 1950, when the Manchester computer had more storage capacity, this was the first question to which Turing turned. This was probably the first time that the power of the universal machine was seriously used to probe the inner life of the numbers.

Fermat had another conjecture – a simpler one – about primes. He studied the numbers which are one *more* than a power of 2. Now, it is (fairly) easy to check that $2^r + 1$ cannot be prime if r has any odd factor. So the only possible candidates for primes are numbers of the form $2^{2^n} + 1$. The first examples are $2^{2^0} + 1 = 3$, $2^{2^1} + 1 = 5$, $2^{2^2} + 1 = 17$, $2^{2^3} + 1 = 257$, and $2^{2^4} + 1 = 65537$. These are prime. Fermat guessed that *all* numbers of the form $2^{2^n} + 1$ would turn out to be prime.

Constance Reid devoted Chapter 7 of *From Zero to Infinity*

to this question. She told the story of how in this case Fermat was proved wrong, because 100 years later Euler showed the next Fermat number, $2^{32} + 1$, could be divided by 641. By 1954 a few further Fermat numbers were known not to be prime. No more Fermat primes have been found, but there is no proof that there are none to be found. In this case, computing has not made much difference.

Fermat did not know that this question was connected with Euclid's classical constructions with straight-edge and compasses. In Chapter 5, we found that drawing a pentagon was tantamount to constructing a line of length $\sqrt{5}$. In Chapter 7, I mentioned the annoying fact that being able to draw a length of $\sqrt{7}$ is not enough to draw a heptagon. The Greek mathematicians knew this, but did not know the whole story. In fact, a regular polygon with *seventeen* sides can indeed be constructed, though in a more complicated way, based on lines of length $\sqrt{17}$. This was shown by Gauss, but even he lacked the complete picture. Only in 1837 was it proved that an n-agon is constructible by Euclid's methods if and only if n has very particular factors. These can include any number of 2's, but at most one of each of the numbers 3, 5, 17, 257, 65537 – and any other Fermat primes, if they exist.

MODERATE: Explain why this means that (in principle) you can construct a 65535 × 65536 × 65537-gon with straight-edge and compasses.

If 65537 is the last Fermat prime, there must be some as yet unknown reason why the remaining infinite number of Fermat numbers are all composite. Or else, there are finitely or infinitely many gigantic Fermat primes. In either case there is something to be found out. The connection of Fermat primes with polygons can be expressed entirely in terms of square roots and squares, so mathematicians certainly don't know everything, even about squaring.

Fermat numbers are probably not of great significance, but this question gives a model, in the way it links numbers, algebra and geometry with other surprising features of advanced mathematics. Certain things do not work, do not fit, except for certain huge and unexpected numbers. The numbers 196883 and 196884 have turned out to give extraordinary connections between string theory, complex numbers and the structures mentioned in Chapter 6 as the 'finite simple groups'. This is where modern research is following Fermat and Gauss and Ramanujan today. These structures involve astronomical numbers, and some people conjecture that they have to do with why the universe has to be so big, and so may restore a pure mathematical basis to it after all. This is certainly an example of how string theory has energised every part of mathematics, including pure number theory – even if it is not, in the end, the key to physical reality. Maybe computers will play a vital role in grasping these connections.

Computers certainly play a central part in breaking the code of the universe: both in cosmology and in particle physics the demands of science constantly push at the boundaries of computing power. (The World Wide Web emerged as a sort of spin-off from CERN, the European centre for fundamental research in physics.) But now we will turn to the breaking of more down-to-earth ciphers.

Hidden patterns

Codes and code-breaking were, behind the scenes, a major factor in encouraging governments to build computers after 1945. The Anglo-American government code-breaking establishments have, ever since, been major employers of mathematicians. Fermat's last prime $2^{2^2} + 1 = 2^{16} + 1 = 65537$ gives a point of departure for illustrating the kind of mathematics that is involved.

I will first use the smaller Fermat number 17 instead of 65537, as a miniature version suitable for a book page. The resulting patterns can then be seen without using a computer. Consider this recipe, short but sweet:

Step 1: Choose any number from 1 to 16.
Step 2: Multiply it by 10.
Step 3: If the result is bigger than 16, subtract 17. Continue subtracting 17 until the result is 16 or less.
Step 4: Write down this result, then go back to step 2.

It produces a sequence of numbers from the following repeating cycle:

. . .1, 10, 15, 14, 4, 6, 9, 5, 16, 7, 2, 3, 13, 11, 8, 12, 1. . .

where the sixteen numbers from 1 to 16 each appear once. The interesting thing about this pattern is that it seems *devoid* of pattern.

The apparently patternless sequence is *not* random, since

each number depends on the previous one in a specific and simple way. The non-randomness shows up on closer inspection: take the first eight values (1, 10, 15, 14, 4, 6, 9, 5), subtract them from 17, and you will find the remaining values in order (16, 7, 2, 3, 13, 11, 8, 12).

In fact a super-intelligent being might say immediately: 'Random? What's random about that sequence? They are obviously the powers of 10 modulo 17!' From this point of view it is no more random than is the sequence 2, 4, 6, 2, 4, 6, 2, 4, 6 . . . which lists the remainders when the powers of 2 are divided by 10 (i.e. the last digits of 2, 4, 8, 16, 32, 64 . . .). It is useful to think in terms of an argument or contest with a higher intelligence, because this brings out the similarity to the problem of *breaking a code* by spotting a pattern in the apparently patternless.

Despite the objection of the higher being, there is a good reason for calling the sequence 'random-looking'. Coding these numbers as binary digits (using 0000 for 16) gives a sequence of 64 bits with very special properties which resemble those of a genuinely random stream, as produced by tossing coins or throwing dice:

0001101011111100100011010010101000001110010001
11101101101001100

EASY: Check that there are equal numbers of zeroes and ones, that zeroes and ones are equally likely to be followed by a zero or a one, and that the sequences 000, 001, 010, etc. are also equally frequent. The frequencies of the various possible sequences of four and five symbols are also close to equality.

MODERATE: Show that powers of 3, 5, 6, 7, 11, 12, 14, also generate random-looking sequences of sixteen numbers, but the remaining numbers do not.

DIFFICULT: The numbers 3, 5, 6, 7, 10, 11, 12, 14, are those which do not appear in the diagonal of the base-17 multiplication table.

So this apparently abstruse investigation of diagonals, the subject of Gauss's Law of Quadratic Reciprocity, leads to something useful: *pseudo-random streams*.

When Fermat's number 65537 is substituted for 17, a longer pseudo-random stream of 65536 × 16 = 1048576 bits, or 128k bytes, is produced. It would take more than the length of this book to write out the bits, which is why the number seventeen is useful to give a miniature illustration of it. Nowadays, a 128k file is typical for a single picture, a mere detail of spam in an ephemeral inbox, but in the early 1980s it was seriously large – the first Macintosh was launched in 1984 with RAM of only 128k. When the first mass-marketed microprocessors became available, a random stream of this length was perfectly adequate for game-playing purposes.

The 'random number' function in the Sinclair Spectrum microcomputer was a program that generated the powers of 75 modulo 65537 as a pseudo-random sequence. The number 75 is suitable because its powers run through all 65536 possibilities, modulo 65537. It is a particularly neat choice because $75 = 2^0 + 2^{0+1} + 2^{0+1+2} + 2^{0+1+2+3}$. This means that only the simplest operations on zeroes and ones are needed to multiply by 75 modulo 65537. (But, as a footnote to the history of computing, I may remark that Sinclair's inefficient code did not even exploit the fact that 65537 is simple in binary representation. Sinclair Research soon vanished from the page of history.)

The well-hidden nature of a pseudo-random pattern makes it suitable for use in ciphers. To see how ciphers are made and broken, however, we should first look at what can be done with truly random streams.

Secret keys

The problem of defining a truly random stream revisits the problem of defining probabilities, as we met in Chapter 6. A puzzling fact is that if you create a random sequence, and then publish it in a book, it is no longer random but very special! One modern idea is that a random stream can be defined as one that can't be produced by a Turing machine any more efficiently than by listing it bit by bit. So pseudo-random streams are definitely *not* random.

Despite this difficulty of theoretical definition, it is not hard to generate a stream which is random for all practical purposes, by using an electronic equivalent of dice-throwing. You can then use such a stream to encipher a message with true security, so that it can only be read by its intended recipient. To do this you use the random stream as a *one-time pad*. First, make just two copies of it; keep one secure and somehow convey the other to your partner in such a way that it cannot be intercepted. Later, you can use this stream of bits

— just once — to send a secure message to your partner. To illustrate how it works, suppose that the message to be sent consists of the four symbols of 'Yes?'. Coding these in standard ASCII, these symbols become 5965733F in hexadecimal, or 01011001011001010111001100111111 in binary digits. Suppose that the one-time pad, as agreed in advance between sender and receiver, consists of the stream 01100010101001001001101001011011. Call these M for message, K for key, respectively. Then the ciphertext C is obtained by combining M and K by addition modulo 2:

```
M  01011001011001010111001100111111
+ K  01100010101001001001101001011011
= C  00111011110000010010100101100100
```

This operation needs only the rules $0 + 0 = 0$, $0 + 1 = 1$, $1 + 1 = 0$. It is equivalent to applying the logical XOR operation from Chapter 2. The ciphertext C can then be safely transmitted and if the recipient of the message applies exactly the same operation with the same pre-agreed key K, the message M will emerge. XOR has the elegant property of being self-inverse: doing and undoing are the same.

This kind of cipher has a counter-intuitive aspect which has been known to scare military minds: half the binary digits of the message are transmitted undisguised. This might be compared with the famous problem of advertising, that half the budget is always wasted. But the total uncertainty about *which* half is unenciphered gives a proof of security which runs like this: assume that the keys really are generated randomly, so that all key-streams are equally likely. Then a cryptanalyst might indeed guess, from the ciphertext as above, that the message is 'Yes!'. But if a rival cryptanalyst guesses that it is 'No!!", there is absolutely no evidence to favour either of these guesses – nor, indeed, to prefer any four-symbol guess over any other.

A truly random key-stream must be as long as the message to be enciphered. This makes it long and inflexible in operation. It also carries the danger of a false sense of security if the pad is used more than once by mistake. This is exactly what happened with certain Soviet messages of the period 1942-1945, which were subsequently broken and became a key element in the Cold War. That said, one-time pads are still perfectly viable for one-to-one systems, and modern personal computers make it easier than ever to manufacture one.

For ciphers used on an industrial scale – as for mobile phone wireless links – you cannot arrange for the physical transfer of CD's. You might therefore be tempted to use a *pseudo-random* sequence instead. Such a sequence has the virtue of being generated by a program and so not needing secure physical transport of huge keys. But its use in encipherment must be quite different. Anyone using a system based on pseudo-randomness must assume the worst-case scenario: that the enemy will find out that it is not random at all. The enemy then only has to find the key, which is generally much shorter than the length of the stream. For the 17-sequence, for instance, the key is just the starting point in the sequence. There are only 64 possibilities, each specified by just 6 bits. For the Spectrum 65537 sequence, similarly, there are only 1048576 keys, each defined by only 20 bits.

The number of possible keys is generally called the *strength of encryption*, but such a stated 'strength' can be highly misleading. As an analogy, a safe may have a PIN combination lock of four decimal digits. This means it has 10000 possible keys, formidable if all combinations must be tried in turn. But a safe-cracker who can detect internal motions in the lock, and so discover one digit at a time, will need only four operations to crack the PIN. Just such a shortcut exists if the Spectrum stream is used. Suppose that you can guess a few letters of plaintext (as was so successfully achieved with the Enigma). A XOR with the ciphertext then produces a segment of the key sequence. This immediately shows the times-75 pattern, identifying the place in the sequence, and so the whole stream. Something a little like this, but much more complicated, happened with the cipher used for Hitler's messages, which was not Enigma-based but used an XOR with a pseudo-random stream. A single fatal message of 30 August 1941 revealed the pseudo-randomness to sharp British eyes.

Even if no plaintext can be guessed, the task of trying out a mere million possibilities (which is what so-called 20-bit security means) would be small beer for a computer now. A trillion keys, 40-bit security, is about the lowest serious number for privacy. This is essentially the basis of the GSM system for mobile phones, which has 54-bit security. It enciphers the handset-to-mast signal with a *shift register* algebra which is an extension and complication of the times-75 idea. When you use a GSM phone you hope that its extended, complicated pseudo-randomness means there is no clever method of eliminating many keys at once – so that there is nothing better than trying out all possible keys, which would take a very long time. The reason why an element of hope is involved is that it is very hard to be *certain* that there is no clever short-cut – for reasons which seem to go deep into the nature of numbers, as we shall see. And in fact, in 2006 it was shown by Israeli analysts that there *is* such a shortcut. The weakness of GSM is that it employs error correction before encryption, and this introduces a redundancy – a guessability – into the transmitted message. Of course, as people often make their mobile phone calls very loudly in public, the attempt at secure encipherment is largely redundant anyway.

Nowadays there is intense, open, international scientific discussion of all such questions, completely different from the 1930s era when Turing took on the Enigma. Cryptology in the scientific sense has little to do with treasure-hunt puzzles popular in fiction and in the media, which require informed guessing of one-off texts using culture-based clues.

Like the computer, scientific cryptology is the outcome of a revolution of the 1940s, in which Turing was a major figure, along with Claude Shannon, the founder of the mathematical theory of communication. In 1944 Turing devised a precursor of GSM: it was a compact speech scrambler using a pseudo-random key, which he called Delilah after the bewitching lady of the Old Testament. It was never used, and like much of Turing's cryptological theory was probably

30 years ahead of its time. The Allied powers-that-were of the Second World War had the advantage of a kind of time travel, as they did also with the enlistment of the world's leading physicists to build the nuclear weapon of 1945. To them, the productions of science were as magical as the powers of Delilah, although they were in fact the exercise of pure reason. But Turing's logical revolution had much further to go.

User-fiendishness

Computers and ciphers, discussed in an open, international scientific forum, have spurred a development that in the 1940s was hardly a dot on the horizon. This is the theory of *computational complexity*. It discusses not whether it is possible or impossible to solve a problem, but *how hard* it is to solve. This question is rather different from the classification of individual Sudoku problems as EASY, DEADLY, or FIENDISH. It measures the complexity of Sudoku as a *whole* by considering how much more difficult Sudoku becomes in going from 3 x 3 to 4 x 4 and on to *n* x *n* puzzles.

The guiding idea is that the difficulty of the problem should be measured by the time taken to solve it by the best possible method. But it is usually far from obvious what the best possible method is. We have touched on this question in Chapter 4 with the finding of a square root, and then in Chapter 6 with the calculation of π based on Ramanujan's out-of-this-world insight. Code-breaking dramatises the seriousness of the question. The story in Chapter 6 of how

Enigma messages could be broken in hours, using Turing's Sudoku-like logic, gives an example of where the existence of a non-obvious fast method was of immense significance to world events.

Just as Constance Reid was explaining binary numbers to the public in 1956, Gödel made a remark addressing this much higher-level question of complexity. Papers began to appear in the 1960s, but only took off after 1970. By that time, computers had made complexity an essential and practical issue in software engineering. Sometimes the provision of faster hardware makes scarcely any difference, compared with the benefits of brilliant software. To measure these benefits independently of the brand of computer, the 1936 concept of the Turing machine became essential. As in other ways, it was only in the 1970s that Turing's work was really appreciated.

As an example of an EASY problem, that of *searching*, or finding needles in a haystack, is one that computers can solve with laid-back unconcern. Not so obvious is that *sorting* files into order is almost as quick as reading them, although the brilliance of search engine software makes this feat vivid on computer screens. With the best method, a million items can be sorted in roughly twenty times the time taken to read them.

In contrast, the problem of finding how best to get from A to B in a network is definitely HARD. (Sending information over the Internet is an example of where this problem would arise.) The London transport system has offered a striking picture of how difficult this is. Until recently,

274

London bus stops had a map showing all the bus routes, and an instruction for planning a bus journey from A to B. You were told to study the routes of the buses passing through A, and those passing through B, and to find a point where they crossed. For the small number of lines on the London underground train system, where the eye is assisted by colour-coding, such a search is feasible. With hundreds of snaking bus routes on a large map, it was generally impossible to perform. It could be done with a pencil, carefully tracing the routes and checking all the possibilities in turn. But it could not be achieved in a single step (let alone in the conditions of a bus stop on a dark night). This impractical bus-map instruction gives a good picture of what is called a *non-deterministic algorithm*. It is something that could be done, but only with an unlimited number of testing and checking operations.

Some people have amazing gifts of visualisation and memory, and it is a tricky question as to what connection such mental powers have with mathematical thought. Often it is said that autistic people, who like following rules and procedures in a literal-minded way, have an affinity with mathematics and science. But that characterisation overlooks the imaginative aspects of mathematical thought, seeing only its precision and rule-following, the things that can be left for computers to do. The charming autobiography of Daniel Tammet, *Born on a Blue Day*, brings out something rather different from this, namely a deep and immediate sensation of the numbers going far beyond One to Nine, and an ability to combine them in extraordinary calculations without consciously worked out steps.

Inspired by his magical picture of the colour and texture of numbers, I want now to imagine an Autistic Angel with the power of being able to perform such a 'non-deterministic algorithm', such as solving the bus-map puzzle in a flash. An Autistic Angel would likewise be able to solve SUPER-FIENDISH Sudoku puzzles in one go, just by seeing the solution and checking its correctness. A third example comes from the problem of primes and factorisation. An Autistic Angel could see in one go whether a number is prime, and find its factors if it is not. As with the bus-map problem, there does exist a method for doing this, if only by trying out all possible factors, but it takes ordinary humans and ordinary computers a long time.

Complexity theorists have defined a class of problems called **P** which consists, roughly speaking, of those that are as quick and easy as searching. The class **NP** then consists of problems which would be quick for an Autistic Angel. It might seem that **NP** is obviously a bigger class than **P**. But no one has yet been able to prove that this is the case. It turns out to be FIENDISHLY difficult to prove the non-existence of an EASY method for an apparently HARD problem. This question is very similar to the problem of ensuring that there isn't a quick clever method for breaking a cipher system.

Resolving this question is another Millennium Prize problem, and so another (HARD!) way to win a million dollars. Nowadays this subject has a vast literature detailing subtle variations and distinctions. Everything suggests that this is a deep problem involving the very nature of symbolic representation. It appears that the apparently abstruse and impractical Gödel theorem described in Chapter 1 has this extension of great practical significance for everything computers do.

Within **NP** there is a class of **NP**-*complete* problems, comprising what are in a sense the hardest nuts to crack: if any of these have easy solutions then all **NP** problems do. Sudoku is claimed as an example of a **NP**-complete problem. The problem of factorising integers is a more important case of an **NP** problem. It is not known to be **NP**-complete, and is thought probably not to be so. But if anyone found a **P** method for factorisation, it would make large waves in the theory of numbers, and also in the practical world. This is because the RSA (Rivest-Shamir-Adelman) system of *public-key cryptography* rests entirely on the apparent DEADLY difficulty of finding factors of large numbers. If factorisation turns out to be EASY after all, the whole system will collapse.

Public keys

The motivation for public-key cryptography must first be explained. Chapter 6 described some basic features of the Enigma cipher machine which made it vulnerable to attack. But a more subtle problem, the *key distribution problem*, is also important. For the Enigma, the key consisted of the starting position of the rotors, plus the plugboard choice: this was the information that the British code-breakers were trying to get, and which had to be kept completely secret. But how would the German sender and receiver *themselves* know the correct key for a message? Typically, one would be a U-boat out in the Atlantic for months, the other a shore station. There are basically two possibilities: (1) to agree the keys in advance, each participant keeping a secure copy, or (2) to communicate the key for the message at the time of the message itself.

As an extreme form of the first option, each could have a physical copy of a one-time pad. GSM mobile phones also rely on the first option, as they depend on a secret key of 54 bits. However, they send, unenciphered, an extra 22 bits which determine the starting point in the key-stream. This is no protection against an enemy who knows the secret key, but makes it very unlikely that the same sequence of key-stream will be used twice. The Enigma systems, in contrast, made extensive and vital use of the second option. The Polish break into Enigma messages before the war was possible because the Luftwaffe used a foolish method for enciphering that part of the key which was made up and transmitted at the time of the message. (The folly lay in repeating a triplet of letters before enciphering them: this primitive approach to error correction added a fatal redundancy which the Polish mathematicians brilliantly exploited. Apparently GSM design has repeated this mistake in a more subtle form.)

E-commerce demands enciphered communication of credit-card details. But the key distribution problem appears to pose an insuperable barrier. Both methods (1) and (2) demand prior contact and agreement between sender and receiver. Retailing is a service to the public, not to a private clientèle, so the retailer can never know who will be buying, nor when. Nothing secret can be pre-arranged.

This problem had been foreseen in the 1970s, a time when the US government was torn between conflicting interests. It was under pressure to underwrite a standard for confidential communication in commercial business. For this, the Data Encryption Standard, as an enormously improved sort of Enigma, was developed. On the other hand, the National Security Agency wanted to keep cryptology completely out of public discussion. Although the conflict continued for decades, the genie got out of the bottle for good. The critical point probably came in 1977 when Martin Gardner published a description of the RSA system in *Scientific American*. A fascinated public then learnt that there is a form of cryptography where the sender and receiver need have no prior contact. It solved the e-commerce problem, twenty years before e-commerce started.

RSA brings out and exploits the difference between the 'it is so' of Platonic existence, and the 'how to do it' approach of Diophantus. The situation could be compared with Sudoku. When a newspaper publishes the puzzle, then 'in principle' it has also defined the solution, which is logically deduced from the information given in the Sudoku square. Yet obtaining that solution is difficult – indeed, the whole point of Sudoku lies in the challenge. In the same way, you can publish a public key. This 'in principle' defines the decipherment method, but in practice, for anyone other than an Autistic Angel, it is a hard, virtually impossible problem. So the idea of public-key cryptography rests on breaking the symmetry of doing and undoing, in fact making it as asymmetric as possible.

The problem of factorising a number with hundreds of digits gives a classic example of such asymmetry. It is easy to multiply, hard to factorise. This problem can also be exploited in a very elegant fashion to give a practical cipher scheme. It makes a recipe based on the tastiest properties of numbers – the powers and primes that have been peppering these pages.

Prime properties

The retailer needs to have two large prime numbers P and Q, and to keep their identity a secret. Their *product* $N = P \times Q$, however, is made public. Multiplying P and Q is EASY, but factorising N is HARD. Everything depends on this broken symmetry: it means that N can be announced publicly and yet P and Q remain effectively secret.

Two other numbers are involved: E and D. The enciphering number E (often 65537) is advertised to the public. But D is the retailer's deadly secret: it is the deciphering number. D is easy to find if you possess the secret P and Q: but otherwise it is as difficult to find as it is to factorise N.

The customer has a message, which we will take to be a large number M. Thinking of a message as a number is natural: any message, written out in ASCII bits, is a long number in binary notation. But M must be less than N. If the message is larger than N, it must be divided into N-sized segments and the following procedure applied as many times as necessary.

The procedure uses modular arithmetic. It runs as follows. Using the publicly available N and E, the customer has her computer work out the number $C = M^E$ modulo N. She transmits C, the enciphered message, to the retailer. Then the retailer, using her secret number D, works out C^D modulo N. The result of this should astonish you, like a mind-reading magician announcing the chosen number that was supposed to be a secret. It will always be equal to the original M.

The customer and retailer each need a computer – one to work out C and one to work out M – but the calculation is not as difficult as it looks from the formulas. There is no need to find the gigantic number M^E, for the work can be arranged so that no numbers bigger than N ever appear. No one but the retailer can get M from C unless they have managed to steal, guess or work out the secret number D.

There are plenty of primes, as already noted in Chapter 6, so there is no danger of running out of suitable P and Q from which to form a public number N, and virtually no chance of two different people choosing the same N. It is also easy to find primes, because it is far easier to test a number for primality than it is to factorise it. The reason for this is that the public-key system itself is very unlikely to work if P and Q are not prime, so trying out a few enciphering and deciphering operations will serve as a good test of primality. This idea can be refined to give a more economical test and causes no practical difficulty at all.

But why does this magic trick work? Why should there exist a deciphering number D which neatly and completely undoes the effect of the enciphering number E? The proof rests on a theory of numbers which goes back to Fermat. It can be considered as an aspect of —

9

End Times Table

The number Nine completes the Sudoku square. Ninth is last: ninth symphonies spell doom. Beethoven had just started on his tenth when he died; Bruckner didn't finish his ninth; and perhaps fearing the same, Mahler didn't call *Das Lied von der Erde* his ninth symphony, with the result that the one he called his ninth, beginning with his failing heartbeat, was indeed the last he completed. But parting has a secret sweetness in its sadness. The Indo-European word for 9 is the *new* number: *novem/novus*, *neun/neu* and *neuf/neuf*. Nine is not the end: it opens the door to the Eurabian *decimals*.

Numbing numbers

Nine is the gateway to the continuous world of measurement,

and the mathematics of functions rather than numbers. Nine opens the way to the serious work of differential equations, and hence calculating climate, quarks, cosmology, space probes, protein shapes and everything else. Unfortunately, the music of Nine does not always resound as stirringly as this vision suggests. Nine is haunted by a tasteless in-store muzak, full of glitches and fudges, with prices like £12.99 trying to look less than they are. Nine, for most people, is not a radiant gateway but a dark door with a lock where the key doesn't work. Nine means sums with decimals, fractions and percentages filling school hours with fidgeting, clock-watching, gloom and anxiety.

G. H. Hardy wrote off school mathematics as useful but boring, adding contemptuously that it was all Hogben understood. This was unfair on Hogben, who must have taken many readers far beyond the scope of school lessons. Constance Reid prudently steered clear of the subject altogether. But her illustration of Zero, that it counts the number of elephants in the room, provokes the thought that this number is actually One. The elephant in the room is the dismal experience of school examinations.

Being boring

In the English system, the 'GCSE' consists of a streamlined form of the Indian-Arab arithmetic introduced to Europe by Fibonacci, plus the decimals introduced in the sixteenth century, together with some snippets of algebra and geometry. It is tested at fifteen or sixteen. The 'A-level' is a simplified and selective version of important seventeenth-century advances, with some extra bits of statistics, and sat at seventeen or eighteen. This pattern continues at degree level, where students start on a more logical basis with topics sorted out in the eighteenth and nineteenth centuries. After three years they reach a selection of twentieth-century advances, and can better appreciate Hardy's planet. By about twenty, it is possible to catch up with current knowledge in one small area – for instance in that heartland of mystery and discovery to which the elliptic functions are the doorway. After that, it is no longer possible to squeeze a century into a year, and you are doing well to discover something new with a doctorate at twenty-four.

It's tough trying to find out new things, for the very few who get to that stage, but I am more concerned here with those first steps which everyone is supposed to have covered. In reality, the sums of medieval Europe are still stumbling blocks for bored teen youth, despite the most dedicated efforts. Perhaps one problem is the very word 'mathematics', and the implied aspiration of embarking on centuries of mind-bending difficulties. It might be better to abandon the concept of teaching 'mathematics', and to concentrate on arithmetic purely as a survival skill. Anyone is at a great disadvantage without numbers for counting, telling the time and date, wages, shopping, credit cards, sharing bills equitably, keeping track of the miserable trail of debts, tax and benefit forms. A reasonable criterion for *mass, compulsory* schooling is that it should cover enough to understand articles, advertisements and graphs in the mass-circulation press. People certainly desperately need this numeracy when faced with baffling bureaucracy, mendacious commerce and 'financial advisers' pushing worthless pension funds and pointless insurance policies.

Whether it is even essential to memorise the base-10 multiplication table, I doubt. It is more important to have a good general sense of what people call ballpark estimates; to know motes from beams, gnats from elephants, sledge-hammers from nuts. Such good sense does not have a handy name, cannot easily be tested, and so is not fetishised like 'times tables'. But understanding what is vital and what is not, making good approximations and good models, and arranging in orders of magnitude, are central ideas in the most serious mathematics, as well as in ordinary life.

After that basic element of force-feeding, I doubt whether any more mathematics should be made compulsory. Notable, successful people say that they see no point to the morsels of algebra and geometry they had to do in school, and that they have never needed them once in their lives. It is nonsense to say that in a technology-based culture, it is necessary to understand science. The broadcast media are dependent on the enormous engineering practice built on the mathematics of Maxwell's equations for electromagnetism, the quantum mechanics of the electron and the concept of the universal machine. But you need know nothing whatever of these to be a star.

In the case of the universal machine, the discussion in Chapter 8 showed how its very user-friendliness makes technical understanding unnecessary. The computer makes an extreme example of what Gauss said about the role of mathematics as both Queen and Servant. The rather aristocratic world of mathematical logic has become the language of – well, servers. In large measure, all the high-level achievements of mathematical understanding likewise carry on as the world's invisible infra-structure. Sometimes it seems too invisible. Square roots were understood in ancient Babylon, but if some economic or statistical formula needs them, it meets ridicule in the press. This gives me pause, if only because the climate change question needs some mathematical appreciation, if it is to be put in the right perspective.

Even so, pushing the unwilling and unmotivated towards this hidden structure is an unappealing aspect of education. Like their brattish offspring, the computer, the numbers are needed to keep the human party going, but they are not good company for everyone. It's not surprising if young people prefer to embark on a more direct exploration of the huge spaces of possibilities in sport, music, alcohol, gang warfare, etc. etc. Brains have evolved to do just that. In contrast, mathematicians have banged their heads on the wall for centuries trying to figure out questions about just a few dimensions.

Lowest common denominator

To limit the compulsory element in education to the most basic number skills needed for everyday use, would be considered a surrender to the 'lowest common denominator'.

But that expression itself tells a story. It is wrongly used to suggest something small and cheap, and this misuse illustrates just how little was actually learned in school in bygone days when more formal arithmetic was supposed to be imparted. A lowest common denominator is generally a rather *grand* number! The term arises in the context of fractions p/q, where p is called a *numerator* and q a *denominator*. To add $1/1 + 1/2 + 1/3 + 1/4 + 1/5 + 1/6 + 1/7 + 1/8 + 1/9$, the trick is to find the *lowest common denominator* from 1, 2, 3, 4, 5, 6, 7, 8, 9. That is the smallest number which can be divided exactly by all of 1, 2, 3, 4, 5, 6, 7, 8, 9 – which turns out to be 2520. You are supposed to use it to write $1/2$ as $1260/2520$, $1/3$ as $840/2520$... $1/9$ as $280/2520$ and so add them up.

In practice, few people add up fractions much more complicated than $1/2$ hour + $1/4$ hour, and I am not surprised that this complicated theory has left little trace on the collective mind. Probably the misuse of the metaphor has arisen by confusing it with the *highest common factor* of a set of numbers. This is typically something *modest*, like my proposal for more realism in compulsory schooling.

Sado-mathematical practice

Asking for realism about schooling is to open up several cans of unwelcome worms. In adult talk, education seems mostly to be dominated by questions about social class, national identity, religious indoctrination, racial groupings and dressing up in uniforms. Further fetish elements, humiliation and punishment, are not far under the surface of 'adult' images of education. For the young it is mainly about finding peer-groups, gaining popularity, coping with bullying and above all – the explosion of sexual desire and everything to do with it. Every novel, film, and soap opera ever made has told a story of hopeless mismatch between official design and actual experience. In 2007 the Education Secretary, already busy pushing 'times tables' (and Jane Austen) on reluctant teenagers, found a new priority in stopping them from putting rude videos on YouTube. Another mammoth in the class-room lies in the fact that intellectual achievements may attract the most relentless aggression, with the word 'gay' increasingly used as the damning verdict on such aspirations.

In this context, the word 'mathematics' should not be invoked as if it implied a world of unchanging, traditional standards. The question of whether school students should do arithmetic without calculators is very telling. It is perfectly true that you must *understand* what you are asking the calculator to do, and be able to interpret what it says. You must know what numbers are all about. If you don't, you end up writing down answers a hundred times too big, or upside down. For this reason it is essential to learn about adding and multiplying without using a calculator. On the other hand, calculators are now available worldwide, as software working on the computers called mobile phones. This marvellous resource should be used to good effect. Gauss enjoyed doing arithmetic in his head, being generally rather unbearable, but other mathematicians of the past would gladly have had a calculator to cut short their long labours on numerical work.

There is no virtue in being made to do repetitive arithmetic in an old-fashioned way, like soldiers drilling on the parade ground for the sake of it.

Submitting to discipline has its attractions. Channel Four television has a good line in bossy gurus for diets and decorating, and running bad-lads boot camps with a back-to-1950s theme. People pay good money to be ordered around by personal trainers, sport trainers, and life coaches. But involuntary servitude is soul-destroying and in mathematics it is the kiss of death, leaving a derelict corpus of half-remembered routines.

There are several mixed-up strands in 'mathematics' education, elements once with sense but now virtually obsolete. One line is that it is vital for trade and industry: roughly Hogben's approach, but now seen through the eye of nothing-can-stop-me-now capitalism. A hundred years ago, even forty years ago, training in commercial arithmetic was needed for an army of accounting clerks in shops, offices and banks. They were supposed to be machine-like calculators – although in reality, little books called 'ready reckoners' were in heavy use. All such mechanical work is now done by calculators or spreadsheet software.

More elevated than these Gradgrind objectives is the proposition that a modern state needs a scientifically trained population to compete successfully. Past panics about Germany and Russia are now recurring with China in mind. There is sense to this aspiration, but science stretches over a gigantic field, of which only a few snippets can get attention in schools. Much of science rests on intense and deep mathematics. It is unrealistically ambitious as the basis of mass education.

Another strand is the element corresponding to Hardy: that of mathematics studied for its own truth and beauty. A hundred years ago, this meant following Euclid's superb logical scheme. This ideal had some sense to it, and Sudoku shows what fascination pure logic has. But in those days Euclid benefited from association with the literary classics of antiquity, then a common culture to the ruling élite, but virtually unknown today. And Euclid's model of logical deduction is anyway now obsolete. In the late nineteenth century mathematicians found deeper foundations in the logic of numbers. While this work has flowered – with the modern computer as one of its offshoots – school geometry has degenerated into pointless applications of a few formulas concerning circles and triangles.

A constructive suggestion

Yet if the dead hand of compulsion were removed, mathematical thinking could become a lively option for those school students who are heading for something scientific or technical. It would suit those who enjoy puzzles and logic games, stimulate those with an eye for the geometry behind music and art, and offer something to those who just want to be different, even *extreme*. But to make it lively I would cut loose from the teenage GCSE syllabus, and in particular, place a new emphasis on using computers.

This does not mean using a computer screen as a medium

for delivering maths-book text with cartoon characters, posing multi-choice quiz questions with beeps and honks, pasting slabs of on-line documents into projects, or buying on-line essays. Nor do I have the content of 'information technology' courses in mind. What I suggest is not dumbing down but wising up: students can take charge of the amazing computer power now available so cheaply, and make it do the donkey work while they think for themselves.

What happened to the kind of logical rigour which used to be taught through Euclid's geometry? It flowed into the logic of numbers, and then through Russell, Gödel and Turing into computer languages. Computers do exactly what you tell them, not what you mean, and writing programs for them reignites the extreme rigour and mind-game challenge that used to go into Euclid. Usually this is thought too hard for school students, and 'information technology' students seem to be trained to parrot waffle about programs (though this may be unfair on parrots). I don't believe it is too hard. The skill of computer gaming is a good step towards programming.

In fact, I suggest going back to the constructivist can-do hands-on philosophy of Diophantus and Al-Khwarizmi. Students don't have to act like machine-slaves, nor pretend to be an above-it-all Athenian aristocracy, but could instead act as the intelligent multi-tasking craftsperson using the staggering resources of a universal machine. The computer can be used for that old-school discipline of logic, *and* for exploring the patterns of number and geometry for their own sake, *and* for acquiring an up-to-date technical and scientific expertise for text, music, photography, graphics and animation. It can be the platform for creativity. Fermat was inspired to a good idea in the margins of a constructivist book, and others may also take off from recipes to cook up new thoughts.

This chapter will look at an old-fashioned school topic: Victorian long division and those ancient HCFs. By actually doing it, and not just talking about it, we will create the number patterns which lie behind the brilliance of the RSA cipher. On the way, we will also ponder on something that goes beyond what computers do: the idea of infinity. For an aural image of this final chapter, hear the Pet Shop Boys end their *Introspective*, suggesting how the music will go on and on and on and on... They end with a huge fading tone-cluster of all the integers, a musical rendition of the ellipsis of the *ad infinitum*. That originality in sound design is also an echo of what mathematics could be like for willing, energetic school students using Alan Turing's wonderful invention.

It's all right...

The *Independent* newspaper recently offered a column about the number 23, to coincide with a film of that name. It didn't mention the plastic number, nor that 23 is a third of *soixante-neuf*, but it did reveal a perfect example of how not to use a calculator:

> ...More freaky numerical coincidences: Charles Darwin's *Origin of Species* was published in 1859 – 1+8+5+9 = 23. Two divided by three makes 0.666 recurring (allegedly – actually it makes 0.6666666667). The Hiroshima bomb was dropped at 8.15am – 8+15 = 23...

Allegedly 0.666 recurring? *Actually* 0.6666666667? The *Independent* writer has put 2/3 into a calculator, pressed the buttons and thoughtlessly copied the display.

It is not difficult to do the division by hand. The raised index figures indicate the *remainders* which arise at each stage.

$$\begin{array}{r} 0.\ 6\ 6\ 6\ 6\ 6\ 6... \\ \hline 3\ \overline{\smash{)}\ 2.^20^20^20^20^20^20^20...} \end{array}$$

3 doesn't go into 2, so write 0 and carry 2; 3 goes into 20 six times, remainder 2, so write 6 and carry 2... and so on, always carrying two and *never ending*. This is what 'recurring' means, and 2/3 is, as correctly alleged by the film, 0.666 recurring. The *Independent's* calculator only stopped at ten decimal places because a calculator is programmed to stop there and display, with the last 6 'rounded up' to 7.

Here are some other results of performing divisions, all recurring decimals:

1/9 = .111111111...
2/9 = .222222222...
3/9 = 1/3 = .333333333... and so on to
8/9 = .888888888...

EASY: Rewrite 1/9 as $1/10 + 1/10^2 + 1/10^3$...
TRICKY: This suggests another way of looking at recurring decimals, using a little algebra. The infinite *geometric progression* $r + r^2 + r^3 + r^4$... can be summed to $r/(1-r)$. Show why this is true by multiplying both sides by $(1-r)$. Then apply it in the case of $r = 1/10$ and show it agrees with the result of the division sum.

The interesting thing about this pattern is that it suggests that $9/9 = 1 = .999999999\ldots$ Is this a contradiction? If you put a price tag of '£9.999 recurring' on a £10 item, would you be breaking the law? We have come full circle, back to the question of the integrity of One. If we turn a microscope on all the fudge and fiddle of Nine, all that seems awkward and doesn't work properly, we can find something new and exact.

The last lap

There is no contradiction. It is true that $1 = 0.999999999\ldots$ recurring. These are different decimal representations of the same number, and this gives us a useful way of looking at the *ad infinitum*.

When running on the treadmill, if you get bored with thinking about relativity, you can think about fractions and infinity instead. There is nothing like an unwelcome duty as a spur to concentrating the mind on rational numbers. When $1/3$ is done, you know that only half as much again is needed to get half way; when $2/3$ is done, only one eighth as much again to get $3/4$ through. As the counter heads towards 10000 metres, the glow of achievement, on seeing $3/4$, $5/6$, $7/8$, and $9/10$ done, burns more and more often. At last, when $99/100$ and $999/1000$ and $9999/10000$ have been done, you experience infinitely many achievements in a flash. That's the problem of '0.9999 recurrring': it seems to mean that infinitely many achievements are needed to express the simple idea of One.

This is a form of 'Zeno's paradox': to run 10000 metres, or indeed one metre, it seems that you have to achieve infinitely many things. This classical problem was addressed by the sorting-out process, which in the nineteenth century made sense of the concept of 'limit', and so of the calculus of continuous curves. Instead of 'going to infinity', and so actually taking infinitely many steps, it is possible to define why $.999999... = 1$ by making precise the sense in which a finite number of steps gets as close to 1 as you choose.

This logic still leaves a *potential* infinity in the concept of 'as close to 1 as you choose', and the infinite sequence of numbers $.9, .99, .999, .9999, .99999...$ still must be considered as existing in some Platonic sense. How this relates to physical measurements, and the world in which we actually take those treadmill steps, is quite unclear. It depends on the truth about the fundamental level of space and matter, the level perhaps defined by strings or twistors. The standard answer, a stopgap reflecting the unfinished business of Chapter 7, is that distances below the Planck distance are not meaningful. But we can leave open this question about how $.99999...$ relates to the physical world, while continuing to explore the inner life of numbers.

Recurring nightmare decimals

The first really strange-looking decimal is for $1/7$ – seven being the awkward customer again. It has an ugly-looking answer. You could get it on a calculator, but that won't show the inner logic which a division sum reveals.

$$\begin{array}{r} 0.\ 1\ 4\ 2\ 8\ 5\ 7\ 1... \\ 7\ \overline{\smash{)}\ 1.^10^30^20^60^40^50^10...} \end{array}$$

Thus, 7 goes 0 times into 1, with remainder 1; carry it forward; 7 goes once into 10, with remainder 3, carry it forward; 7 goes 4 times into 30, remainder 2, and so on. At the last stage, we have 7 going 7 times into 50, with a remainder 1. After this we can write '...' because we are back to where we started: the decimal recurs. So

$$1/7 = 0.142857142857142857142857...$$

has a recurring cycle of six digits.

To calculate the decimal for $2/7$, it is enough to notice that it starts off with a remainder 2. After that, we have already done the work in the division sum for $1/7$: we just pick up the calculation where the remainder 2 occurs, and read off the answer as $0.285714...$ recurring. I will call this the *carousel* property, thinking of the baggage circling after it has emerged from the chute in the airport arrivals hall. Wherever you stand, the same items pass by. To find the decimal for $3/7$, just stand at the point where the remainder 3 occurs, and then watch the cycle thereafter appear as $0.428571...$

There are only six possible remainders: 1, 2, 3, 4, 5, 6, and and each of them arises just once. This means that the number 142857 has the carousel property

$$142857 \times 1 = 142857,$$

$$142857 \times 2 = 285714,$$
$$142857 \times 3 = 428571,$$
$$142857 \times 4 = 571428,$$
$$142857 \times 5 = 714285,$$
$$142857 \times 6 = 857142.$$

Now a connection with nines emerges:

$$142857 \times 7 = 999999.$$

The remainder is 1 when six figures of $1/7$ have been found. So 7 divides into 1000000 with remainder 1, so 7 divides exactly into 999999, 142857 times. Alternatively, you can start with $7/7 = 0.999999...$ and divide both sides by 7; you will get exactly the same answer for $1/7$, but this time getting a remainder 0 after six 9's, instead of remainder 1 after six 0's.

Either way, the fraction $1/7$ and its decimal tell a story about integers: seven divides exactly into the number written as six nines.

EASY: The decimal expression is equivalent to the geometric progression $1/7 = 142857 \times (10^{-6} + 10^{-12} + 10^{-18}...)$

From now on we will leave aside the infinite addition, and concentrate on the finite numbers like 142857 which emerge on the carousel. Is there a pattern which extends to other fractions?

A long road to ruin

We shall find that there is indeed a simple common pattern

for $1/p$, if p is a prime. But if p is greater than 10, knowing the multiplication table is not enough to do the division sum for $1/p$. For $1/17$, you need to *pre-compute* an extra table of information: $1 \times 17 = 17$, $2 \times 17 = 34 \ldots 9 \times 17 = 153$. You then use this to execute the same logic as for $1/7$, but with the remainders worked out by subtraction sums.

This is *long division*. It has been dropped from school maths as unnecessary. My attention was drawn to this by a letter in the *Guardian* to the effect that the kind of people clever enough to do long division are the kind of people who build weapons of mass destruction. This is a valid criticism of mathematics, which Hardy might well have agreed with. Maybe a wise old Babylonian foresaw the trouble that long division would bring in its train. Alas, the Second Law prevents the unspilling of milk. Perhaps more fundamentally, the invention of agriculture was the disaster; since then it has been downhill all the way as the long division of land into property started the cycle of tribal war and mutual destruction. If humanity had remained as tropical hunter-gatherers, like the Pirahã with their blissful ignorance of abstractions, then none of these problems would have arisen, and there would be no letters to the *Guardian* either.

You can ask a calculator for $1/17$ but it probably won't display enough digits to show the pattern. So it's still useful to be old-fashioned and do $1/17$ with a strict, firm hand:

```
          .0588235294117647...
    17 | 1.0000000000000000...
          85
         150
         136
         140
         136
          40
          34
          60
          51
          90
          85
          50
          34
         160
         153
          70
          68
          20
          17
          30
          17
         130
         119
         110
         102
          80
          68
         120
         119
           1
```

The steps begin: 17 goes 0 times into 1, remainder 1; 17 goes 0 times into 10, remainder 10; 17 goes 5 times into 100, using $17 \times 5 = 85$, and working out the remainder 15 by subtraction. 17 goes 8 times into 150, using $17 \times 8 = 136$...

After a while, you find yourself working like a machine to carry out these steps. There is little virtue in doing this once you have got the idea. To learn about the real life of numbers, it would be much more valuable to program a computer to perform the routine and so discover the repeating cycle.

The remainders run in a repeating cycle of 16: (1, 10, 15, 14, 4, 6, 9, 5, 16, 7, 2, 3, 13, 11, 8, 12) and 1/17 = 0.0588235294117647... has a recurring cycle of length 16.

The sixteen-digit number 0588235294117647 also has the carousel property, and without any more work:

$$2/17 = 0.1176470588235294...,$$
$$3/17 = 0.1764705882352941...,$$
$$4/17 = 0.2352941176470588... \text{ and so on.}$$

Equivalently, 17 divides exactly into 9999999999999999 = $10^{16} - 1$. A pattern emerges: 16 is just one less than 17, just as 6 was one less than 7. Does this hold for every prime? For example, does 13 divide exactly into $10^{12} - 1$?

The situation with $1/13 = 0.076923076923...$ is actually a little different. The number 076923 has the rotating property when it is multiplied by 3, 4, 9, 10 or 12; but when multiplied by 2, 5, 6, 7, 8 or 11 it gives a different sequence: 153846. This is because the remainders break down into two separate cycles: 1-10-9-12-3-4 and 2-7-5-11-6-8. But none of this affects the main feature. Both cycles are of length 6. and this is just half of 12. So the decimal for 1/13 does repeat after 12 digits, and it is likewise true that 13 divides exactly into 999999999999 = $10^{12} - 1$.

This pattern will always hold. The decimal for $1/p$ will always return to its starting point in $(p - 1)$ steps, though it will not always return *for the first time* in $(p - 1)$ steps. It is worth experimenting with numbers and then seeing why the decimals logically must show the patterns that emerge.

EASY: 1/11 has cycles of length 2; 1/37 has cycles of length 3, 1/101 has cycles of length 4, 1/41 has cycles of length 5. Check these fit the rule.

HARD: In working out the decimals for any $1/p$, every remainder has a unique successor and also a unique predecessor. HARD: Each cycle of remainders must be of the same length, and that length divides $(p - 1)$.

We draw the conclusion that if p is a prime other than 2 or 5, p divides into $10^{p-1} - 1$. Equivalently, 10^{p-1} is congruent to 1 (modulo p).

But there is nothing special about the number base 10. It is equally true that if p is a prime and m is any number greater than 1, not a multiple of p, then p divides exactly into $m^{p-1} - 1$.

This fact is known as *Fermat's Little Theorem*, because Fermat stated it in 1640, though without a proof. Euler published a proof 100 years later, considerably extending it. There are quicker proofs, but exploring decimals gives a picture of Fermat's discovery as a pattern of interweaving numbers.

Beyond Fermat

The RSA property needs just a little beyond Fermat's Little Theorem: not going as far as Euler's contribution, but just enough to double it up. We need the fact that if p and q are both primes, and m a number which is not a multiple of p or q, then $m^{(p-1) \times (q-1)} - 1$ is divisible by $p \times q$. This is equivalent to saying that the decimal for $1/(p \times q)$ repeats at the period length $(p - 1) \times (q - 1)$. This can be seen by the following argument.

First, we need the fact that if m is congruent to 1 modulo n, then so is any power of m. (This is like seeing why all the powers of 11 end in a 1.) Next take two primes, p and q and a number m which is not divisible by p or by q.

Then Fermat tells us that m^{p-1} is congruent to 1 (modulo p). So its power $m^{(p-1)\times(q-1)}$ is also congruent to 1 (modulo p).

Also $m^{(p-1)\times(q-1)}$ is congruent to 1 (modulo q) by the same argument.

So $m^{(p-1)\times(q-1)} - 1$ is divisible by p and divisible by q, so by $p \times q$. If k is any integer, $m^{k\times(p-1)\times(q-1)} - 1$ is also divisible by $p \times q$.

The RSA cipher rests on this fact. Suppose we have two numbers d and e such that $d \times e$ is congruent to 1 modulo $(p-1)\times(q-1)$.

Then $d \times e = k \times (p-1) \times (q-1) + 1$ for some integer k.

Then $m^{d\times e-1} - 1$ is divisible by $p \times q$.

Multiplying by m shows that $m^{d\times e} - m$ is divisible by $p \times q$, i.e. $m^{d\times e} = m$ (modulo $p \times q$).

This final statement is still true even if m is a multiple of p or q, so we can forget the tiresome restriction on m.

Now think of the number m as the message number M, e as the enciphering number E, d as the deciphering number D, and n as the public-key number N.

We have $M^{D\times E} = M$ (modulo N). This means that taking the Dth power always undoes the effect of taking the Eth power.

So if $M^E = C$ (modulo N), $C^D = M$ (modulo N), just as required for the RSA cryptosystem.

Highest common factor

There is one thing left. The number E is the enciphering number, made public; and D is the secret deciphering number. But given E, P and Q, how do we find a D such that $D \times E = k \times (P - 1) \times (Q - 1) + 1$?

This does not need advanced modern mathematics. It does not even need Fermat's insights. It needs something from Euclid: the ancient Greek algorithm for finding the *highest common factor* (HCF) of two numbers. Our technical exploration of numbers can end in some constructivist mathematics from the heart of classical antiquity.

The construction can be shown by example. Take $N = 77$, so $P = 11$, $Q = 7$, and $(P - 1) \times (Q - 1) = 60$. Given an enciphering number E, we need to find a number D such that $D \times E$ is one more than a multiple of 60.

There is no hope of finding such a D if E has a common factor with 60. For instance, if E is even then any $D \times E$ must be even, and so cannot possibly be 1 more than a multiple of 60. The HCF of E and 60 must be 1. This is a necessary condition, but it turns out also to be sufficient. Given an E, we use Euclid's algorithm to check that the highest common factor it shares with 60 is 1. Then, if it is, we can run the algorithm backwards to find what D is.

As an example, take $E = 43$, so that we must examine the pair of numbers (60, 43) for common factors. One way of doing this would be to find all the factors of both numbers and then check them against each other. But Euclid's method is much simpler and smarter. It works by repeated division and taking remainders. Each time, it reduces the pair to a smaller pair of numbers which have the same highest common factor. Eventually it must reduce to zero, and then the HCF is found.

Start with $(60, 43)$ then write:

$60 = 1 \times 43 + $ remainder 17. This reduces the pair to $(43,17)$.

$43 = 2 \times 17 + $ remainder 9, reducing the pair to $(17,9)$.

$17 = 1 \times 9 + $ remainder 8, reducing the pair to $(9,8)$.

$9 = 1 \times 8 + $ remainder 1, reducing the pair to $(8,1)$.

Now 8 and 1 have highest common factor 1 and that's the end.

To find D, we work backwards through this logic, turning each line into a statement about 1:

$1 = 9 - 8$.

$1 = 9 - (17 - 9) = 2 \times 9 - 17$.

$1 = 2 \times (43 - 2 \times 17) - 17 = 2 \times 43 - 5 \times 17$.

$1 = 2 \times 43 - 5 \times (60 - 43) = 7 \times 43 - 5 \times 60$. So $7 \times 43 = 1$ (modulo 60), so $D = 7$.

Provided that the HCF is indeed 1, this will always result in a unique deciphering number D. It is an EASY algorithm which computers can perform in a flash even for 100-digit numbers. There is a connection with Fibonacci numbers: two neighbouring Fibonacci numbers are the *very slowest* to reduce and confess their lack of any common factor but 1. That is the sense in which they are the most unneighbourly numbers of all, the property mentioned in Chapter 5.

EASY: Find the HCF of 75025 and 46368 by this method.

Using RSA encipherment

Actually working out an RSA encipherment makes a TRICKY puzzle. Again, take $N = 77$ and $E = 43$. You can encipher the message $M = 51$ as follows. Use a calculator to show that modulo 77, $M^2 = 60$, $M^4 = 58$, $M^8 = 53$, $M^{16} = 37$, $M^{32} = 60$. Use $43 = 32 + 8 + 2 + 1$ to show $M^{43} = 2$, so $C = 2$. Check that $D = 7$ does indeed effect the decipherment because $C^7 = 51 = M$.

These calculations only illustrate the principle of how the D is found, and how the D undoes the E. They are of no use whatever as a practical cipher, since the prime factors of 77 are obvious. The whole point of RSA is to use a huge N which cannot in practice be factorised. For actual encipherment using such numbers an ordinary calculator is useless: you need software such as `Maple` or `Mathematica` which can handle large numbers. Even better, you can learn a lot from writing a program yourself to handle such large numbers. It is then quite magical to see it reproduce the original message M.

With a public-key cipher, you cannot decipher something you have yourself enciphered. Suppose, for instance, that you were an industrial spy transmitting data to China, and had your computer encode it by RSA without even seeing it yourself. Then you would not be able to decipher the messages that you yourself had sent. If you were caught by the British government and ordered to decipher your messages, in accordance with its Regulation of Investigatory Powers Act, you would not be able to obey.

As with many other computer applications, this is probably not quite what Lancelot Hogben had in mind when promoting the usefulness of Mathematics for the Million. Nor is it what Hardy expected. In 1940 he wrote that 'no one has yet discovered any warlike purpose to be served by the theory of numbers or relativity, and it seems unlikely that anyone will do for many years'. By 1945 everyone knew that $E = mc^2$ had an application; in 1946 Turing reported on the military importance of combinatorial problems, as an aspect of future computer power. The RSA cipher is just the tip of an iceberg of crypto-mathematics, and the superpowers have armed themselves with the theory of numbers. One key development, which neither Hardy nor Hogben foresaw, is the speed and cheapness of electronic computing.

The relation between theoretical and practical has changed so much in the past 60 years, that it is unwise to predict how the picture may change in the future. There is certainly a possibility of more efficient factorisation methods being found, and so of the whole system being rendered useless. Current methods are mostly still based on refining an observation of Fermat, that the difference between two squares always factorises. This is a point where very advanced developments in the theory of numbers can combine with the fantastic power of computing. A new Euler or Gauss, Riemann or Ramanujan may find something completely new.

Another possibility is that *quantum computing* will make

factorisation into an EASY problem to solve. Theory already shows that this is true: there is a way of exploiting the Two-ness of quantum mechanics to find factors more quickly than by any computer method so far known. It is an odd fact that the problem which shows the greatest theoretical promise for quantum computing is also the one of most commercial relevance. It is not yet feasible to build quantum computers with enough storage to factorise seriously large numbers. But there have never been so many new minds at work with new ideas.

Down to earth

The RSA algorithm requires rather intensive calculation and in practice it is not used for the entire communication of, for instance, a secure webpage. It is used only to solve the key distribution problem: i.e. to transmit the secret key for a conventional cipher system on which the bulk of the message is sent. But that could still be broken by a cryptanalyst able to search through all the possible keys for that system. A special lock on your front door is of no use if anyone can open a window instead.

Another important limitation of RSA is that anyone who *guesses* the exact message M can verify it simply by encoding M with the public E and N, and checking that it does indeed agree with C. A computer could do this checking process for a million, or a billion guesses. So for true security it is not enough to encode a message M: a 'padded' M must be created by adding random material that cannot be guessed. But ensuring that this padding is truly random is itself a serious cryptographic problem.

In practice it may be more important that the secret deciphering number has to be kept on a hard disk, vulnerable to spyware smuggled in by cookies, hidden cams and burglary. An even more down-to-earth point regarding e-commerce on the Internet is that there is little point to elaborate mathematical encryption when the data may be collected and sold by people working in remote call centres. Or when, indeed, people are so easily taken in by scams that they cheerfully divulge their secret data to complete strangers.

These considerations are salutary reminders of the place of mathematics in the world. Human minds can mess everything up. But first, a little more about what human minds have realised.

Numbers beyond computers

Long division reveals a lot about the fractions, or rational numbers. Written as decimals, they are exactly those that, after a certain point, repeat in a cycle forever. What about the other, *irrational*, numbers? We have met $\sqrt{2}$, $\sqrt{5}$, e and π, amongst others. It is quite common to hear it said that π 'can't be found exactly'.

This is true in the obvious sense that π needs an infinite decimal. But then, so does the decimal for $1/7 = 0.142857142857...$ The symbols '...' or 'recurring' are actually *rules*. In this sense, $\sqrt{2}$, $\sqrt{5}$, e and π are no different, or are different only in degree, from 1/7: there are *rules* for finding the entire decimal which are *more complicated* than saying 'repeat the cycle for ever', but are still finite rules, expressible as computer programs. This makes them computable numbers: the title of Turing's 1936 paper was 'On *computable numbers*, with an application to the Entscheidungsproblem'. Computability answers the question of what π is, *exactly*. Its decimal, as everyone who studies it remarks, looks competely random and patternless. But actually it is *pseudo-random*, generated by rules which go as deep as Ramanujan into the structure of number.

How many numbers are there in the continuum? Obviously there are infinitely many. Not so obviously, this infinity is infinitely greater than the infinity of the integers! This comparison can be made exact. It comes from seeing that a real number is equivalent to a *set of integers*. For instance, the fractional part of the number π, written in base 2, is: .00100100 00111111 01101010 10001000 10000101 10100011 00001000 11010011... Knowing this is equivalent to knowing the infinite set of numbers (3, 6, 11, 12, 13, 14, 15, 16, 18, 19, 21, 23, 25, 29, 33, 38, 40, 41, 43, 47, 48, 53, 57, 58, 60, 63, 64...) which mark where a 1 appears. Conversely, any set of integers, finite or infinite, defines a number between 0 and 1. We are using the either/or logic of the number Two to do this.

TRICKY: Which numbers are represented by two sets of integers? How many such numbers are there?

Given a set of N things, how many subsets does it have? The subsets can be counted like this: the first element may be in or out, giving two choices, the second element likewise, giving two choices, and so on to the Nth element, giving N choices altogether and so 2^N possibilities. Thus there are 2^{64} choices for the subsets of 64 items, corresponding to the 2^{64} different possibilities for a 64-bit register in a computer chip. This fact gives a more primitive picture of what a 'power' is: a counting of subsets. And it can be extended to *infinite* sets, for instance the set of integers. If we say there are ∞ integers, we can say that there are 2^∞ sets of integers. This is the step that Georg Cantor took in the 1870s and it lit a long fuse which did not explode until the twentieth century. Cantor's proof that 2^∞ is definitely a greater infinity than the ∞ of the integers, turned into the argument which shows there is no set of all sets. It led on to Russell's efforts to repair the problem and thence to Gödel and to Turing. This, the basic idea of set theory, is where we came in with Chapter 1! Incidentally, the well-intentioned efforts to bring 'modern maths' into the school syllabus with 'Venn diagrams' for sets are not, in fact, addressing the serious content of modernity at all. Those diagrams simply sharpen appreciation of AND and OR. It is only in the study of *infinity* that set theory matters.

In contrast, there are no more rational numbers than there are integers. This is because the rationals can be put in a kind of alphabetical order, starting with the simplest numerators and denominators. The same applies to computable numbers, given Turing's clear definition. This is because each computable number comes from a program, and like recipes in a cookery book, programs can be put into alphabetical order.

There are infinitely more real numbers, defined as decimals with infinitely many digits, than computable numbers. In fact the computable numbers are infinitely sparse: a random real number has a zero probability of being rational and a zero probability even of being computable. Almost all real numbers are uncomputable; all the thickness and solidity of the continuum comes from numbers which have no effective definition. So in what sense do they exist? At this point mathematicians can take different views, and explore different levels of logical axioms about what to assume, a topic that is meat and drink to logicians. There are quite marvellous discoveries about ways of looking at the continuum: 'p-adic numbers' as a completely different picture of how the real numbers fit together, and 'hyper-real numbers', allowing for infinitely large and small numbers. From Chapter 8 you will guess that exponentiation of sets can go on to really really large sets of size $2^{2^{\infty}}$ – and far beyond. Rudy Rucker's book *Infinity and the Mind* takes up the challenge of explaining what happens.

Coming down nearer to earth, a vivid picture of the infinite inner life of the integers has become accessible in the spectacular computer-generated images of the famous *Mandelbrot set*. It sums up much that has happened since Constance Reid's 1956. It is a two-dimensional space of complex numbers, made tantalisingly visual. In fact it brings alive the idea of a space of possibilities, leading to a picture of chaos.

Although the basic features were understood in the late nineteenth century, they could never have been found in such extraordinary detail without computer power. Your own computer experiment can find out how its solid blobs correspond to integers and how their meeting expresses multiplicative structure. (The biggest blob, the heart-shaped cardioid, corresponds to 1. The circular disc to its left corresponds to 2.) But its infinite intricacy poses other questions which cannot be answered by computers, and still defeat mathematical minds.

Introspective...

It is worth thinking about why algorithms are so central. The reason is to do with large numbers: an algorithm is needed when there are more possibilities to deal with than could ever be listed as a table. Politicians always say they won't answer hypothetical questions, although they pass laws to deal with hypothetical situations. Programs have no such luxury of choice; they must deal with input that has never been foreseen. The number of *potential* images, or text files, or webpages, which a program must be able to cope with, is far more than a googol, far more than astronomical. Typically, a program embodies a definite principle, or finite set of principles, applicable to an effectively infinite number of possible cases.

Can the brain do any better? While working on the first computers after 1945, Turing became very keen on the idea that everything the brain does must be an algorithm, from which it follows that the computer could do it too. Turing devised his famous Test to put this comparison of computer and brain on a reasonably objective footing. This is where Roger Penrose, in *The Emperor's New Mind* and *Shadows of the Mind*, takes issue. He makes an argument from Turing's original 1936 work to the effect that human minds can see the truth of something that no program can show. This is because we know what numbers are, but the computer only has symbols. A program that prints out .999999… for ever knows nothing about why that number is the same as One, or indeed what One is. It is like a cook that doesn't know what food is, but obediently prepares it in accordance with orders. Penrose's argument is central to the question of human consciousness, freewill and responsibility – and of course it agrees with what everyone feels: 'I think, therefore I am not a machine.'

Turing was highly alive to these objections, even though he considered that they should be overruled, and it is a paradox that he of all people had an acute sense of individuality and freedom. I have said that he was affected by the climate of thought about mathematics and logic in the 1930s, but he was also driven by an intense and emotional interest in the nature of the mind. This becomes clear from his personal writing in about 1932, which I was able to bring out in *Alan Turing: the Enigma*. The human factor, at that stage, led him to think that consciousness must have something quite definite to do with quantum mechanics. Penrose's conclusions are remarkably similar. Turing himself went back to physics in 1953, with a particular interest in the mysterious process of measurement that turns complex wave-functions into real probabilities.

Logic and physics were each transformed in the revolutionary period of a century ago. But science and mathematics tend to divide into non-communicating sectors, and the explosion of knowledge in the last 100 years has created almost impossible barriers. Thus the question of what 'one photon' means seems to attract little interest from logicians, although quantum information, quantum computing, the quantum Zeno effect, interaction-free measurement, and teleporting pose this question with new acuteness. It is hard to believe that another 100 years will pass without a major rethink of the integrity of One, the continuity of Nine, maybe the supersymmetry of Two and the space-time reality of Four.

But as ever, a question mark surrounds that word 'will'.

Back to square one

The sequence 1-4-9 is the Star Gate in Arthur C. Clarke's

2001, but the events of the real 2001 came as quite a surprise. The human factor is unpredictable. The Intergovernmental Panel on Climate Change makes this unpredictability strikingly vivid. There is a wide spread of outcomes for anthropic climate change depending on what the *anthropos*, or at least its rulers, chooses to do with the Anthropocene geological era it has started. As *New Scientist* comment puts it, there are tipping points in political choices as well as in physical evolution. And as scientists know, doing the computation itself affects those choices. They may express caution in predictions, sensing the danger of crying wolf through premature announcements. But they may actually have been overcautious: revisions to the model so far have included more slippery ice-sheets and glaciers, greater methane release, less oceanic absorption of CO_2, and more trouble from uncontrolled logging, than expected. There seems to be a Gödelian impossibility about finding the correct balance, one which encourages neither complacency nor fatalistic pessimism.

Mathematical prediction comes up against the mystery of human consciousness. There is a parallel in cryptology. Turing and Shannon created the theory of mathematical information as a development of probability and statistics, by using letter frequencies as a start. Such frequencies can be turned into numbers. But letters join into words, and words shade into meaning, sense and truth. Information slides into the human purpose of the messages; this cannot be quantified, and depends on knowledge and understanding. This is not an abstract philosophical distinction. The strength of the British code-breaking work (after a classic Two Cultures mismatch in 1940) lay in the synthesis of mathematical theory and human understanding. There may be parallel questions now. The sophisticated mathematical work at NSA/GCHQ is of little use if no one can understand deciphered messages because all the Arabic linguists have been sacked for being gay, or if their content goes to a faith-based circle in which the 'reality community' is denigrated.

Alan Turing was himself a serious loss to Anglospheric intelligence-gathering. On a notable day, 6 June 1944, he reported on his Delilah speech encipherment, decades ahead of GSM, but never used. His computer plans were stalled, his software plans never used. When he was revealed as a gay man, his work for the state was finished; on 7 June 1954 he was dead. Oppenheimer was widely denounced as disloyal that same week. Once people find themselves swimming against the human tide, molehills of dispute become mountains of contentiousness and even great scientists become as welcome as moles. Discovering black holes, inventing the computer, count for nothing. And like Cold War science in the 1950s, climate predictions have had to face a cool reception – notably in the United States, but more generally against a global tide of commitment to unstoppable growth requiring minimal state regulation of economic activity.

Reason is a back-seat driver, with responsibility and no power, unwelcome on board, unable to strike, and too often sniped at by other, weightier passengers for being a mere materialist. Nor is reason a single voice, but a dialectic, an argument taking centuries and never really finished. Is it cool to do serious mathematics? I am – well, cool about it, as cool as Cordelia was to Lear. It is an enormous privilege to be detached from the torture and slavery under which most of the world labours. But it does not have much to do with optimising personal profit or pleasure as commonly understood. Mathematics needs a stoicism – a G. H. Hardiness, and a Thomas Hardiness too – which runs at odds with noughth-impression, have-it-now, sound-bite culture. Hardy did not like Hogben describing serious mathematics as having a 'coldly impersonal attraction', but cold and impersonal it is, though some like it cold.

Turing's own summary in 1948 of a mathematician's life was that 'he may perhaps do a little research of his own, and make a very few discoveries which are passed on to other men'. Though a supreme individualist, he cast his role as being in a long process of collective advance. As individuals, some come out on top, but many more must lose, and do so gracefully. And mathematics can be cruel rather than cool to those individuals. Gödel's 1931 work made a point of showing that Russell's program had missed the point; while Russell himself had dropped a similar bombshell on an earlier logician, Frege.

Mathematics sometimes tries to re-brand itself and court popularity by advertising fame and fortune, Botoxing those lines that betray hard thinking. But if fame and fortune are the priorities then there are far more enjoyable ways of achieving them. The PSB get it right about Opportunities: 'Doctored in mathematics, I could have been a don, I can program a computer...' – but sensibly, they took the opportunity to make music. Few make the journey in the other direction. Brian May has shown it is possible, finishing a doctorate in astronomy after 30 years, but perhaps only someone capable of performing with Freddie Mercury could rise to that challenge.

Numbers are cool only in the sense that the coolest people can ignore fashion. Numbers are cool in being adult, and showing naked reality. Numbers have the icy-cool taste of eternity, the four-dimensional view. Writing this book, and so being taken back to Constance Reid's book of 50 years ago, has refreshed that timelessness. Perhaps the sweetest thing about One is the glorious first step. As we reach the ending, Nine can remember the music of primal vision, even though it bears a weary experience of the world.

The *anthropos* is faced with the outcome of its own dissipated youth, now suddenly having to grow up while keeping cool. It looks like Mission Impossible, but the RSA cryptosystem is a model of how cool heads can overcome an apparent impossibility. That apparently deadly either/or can be transcended – by using mathematics to analyse its own hardness. Having solved one problem, new questions open up. The entropy crisis may have solutions using new methods: they will need new mathematics, and will lead to new problems. Adult mathematics is the only hope, even if it increases the creases in the human face.

From OSX to infinity

Why does the planet work in operating system Ten?

Mathematicians usually disparage base-ten notation, with its tiresome over-large, over-complicated multiplication table. It seems an arbitrary choice driven by the ten biological digits. But that product of primes, two and five, is a good reminder of our complexity. The Two has the basic broken symmetries of time and space. Doubling and redoubling to calculate with fours and eights would have been highly utilitarian, but the Five has an added and magic aesthetic of imagination. An ever-growing sphere of very faint radio waves, with the radius of a hundred light years, now surrounds the planet. It carries those human two-times-fives. If alien life is curious, or hungry, it can figure them out for itself.

Yet Ten has a life of its own, nothing to do with sticky human fingers grabbing quick fixes. It is connected with the four-squareness of the physical world. Ten is $1 + 2 + 3 + 4$, and that triangular number counts Einstein's equations for four-dimensional curved space-time. Those equations, according to present and imperfect understanding, don't seem to care whether *anthropos* is around to think about them or not. The future, on its grandest scale, is determined by the properties of –

Hands-on Experience

To explore the numbers One to Nine further, you may like to try these extra puzzles. The first set of these go further into the topics of Chapter 9. They probe the inner life of numbers that lies beneath the surface of a calculator.

GENTLE: Check that 11 divides into $2^{10} - 1 = 1023$, $3^{10} - 1 = 59048$...

MODERATE: Find the factors of 999999 and so show that 3, 7, 11, 13, 37 are the only primes p such that the decimals for $1/p$ recur in cycles of 1, 2, 3 or 6. Which primes p have decimals repeating in cycles of 4?

HARD: Which primes p have decimals for $1/p$ repeating in cycles of 7?

EASY: Check that the remainders that arise in working out $1/17$ give the same sequence as the pseudo-random sequence in Chapter 8.

HARD: Show why $1/5$ is 0.001100110011... in binary notation. A calculator supplied as software on a computer may be working directly with binary numbers, and thus have the feature that it cannot store $1/5$ exactly. Alternatively, it may be designed to work as if it was storing numbers in base-10. Power Macs of the 1990s did the former; the Mac OSX does the latter. Experiment to see what your computer does. Ask for $1/5$ and then do repeated multiplication by 2 or 10 to find out what it is actually storing. How does it store 1/3 or 1/7?

TRICKY: The fact that the decimal for 1/13 breaks into two cycles of 6 is equivalent to the fact that 36 and 49 are congruent to 10 modulo 13, so that 10 appears in the diagonal of the modulo-13 multiplication table. Why does this show that 1/13 has a recurring full-length cycle of 12 when expressed in base 6 or base 7?

DEADLY: 1/89 begins with 0.011235... and this points to another appearance of the Fibonacci numbers. 1/89 is actually equal to $0/10 + 1/100 + 1/1000 + 2/10000 + 3/100000 + 5/1000000 + 8/10000000 + 13/100000000...$ Show why this is true by multiplying this sum by 89 (hint: $89 = 100 - 10 - 1$). Without doing a division sum, see how the decimal for 1/9899 begins.

TOUGH: For a quicker proof of Fermat's Little Theorem, look at the pth row of the Pascal triangle from Chapter 6, for a prime p. All the entries other than the two 1's are divisible by p. Why is this? So $2^p - 2$ is divisible by p. As a slightly harder problem, find a way to extend this result to $m^p - m$.

As you may have noticed, not everything in this book has to be taken completely seriously. You have to judge this for yourself. Here is a last puzzle which poses a similar challenge, wrapping together logic, numbers and some curiosities from life and art. The answers are all numbers from One to Nine that fit into a Sudoku. There are twenty-six clues, and you must solve most but not all of them to complete the Sudoku. Many of them make use of modular arithmetic. Some are quite straightforward, but some are as tricky as code-breaking. If you can solve them all, you can count yourself a true expert in counting from One to Nine!

A		B		C				
		D	E	F		G		H
	I					J		K
						L		J
M		N	O			P	Q	R
			S					
N	T			U		K		V
		W	M				T	
X		Y		O	Z	E		G

A: Find the last digit of $9^{8^{7^6}}$.

B: Take the sequence 123456789. How many permutations leave exactly four numbers unchanged? Use the penultimate digit of your answer.

C: Find the penultimate digit of the trillionth power of 2.

D: How many anagrams are there of ONETONINE? Use the sum of the middle three digits.

E: Find the 666th digit, after the decimal point, of 1/666.

F: Six women and six men are at a party. In how many ways can they be arranged into man-woman pairs? Use the second digit of your answer.

G: At the same party, in how many ways can they be arranged into man-man and woman-woman pairs? Use the last digit of your answer.

H: At the same party, in how many ways can the people be arranged into pairs irrespective of gender? Use the penultimate digit.

I: Take the sequence 123456789. How many permutations leave exactly three numbers unchanged? Use the third digit of your answer.

J: How many anagrams are there of ANAGRAM? Use the penultimate digit.

K: Assuming that the world was created in 4004BCE, in which year was it 6000 years old? Use the last digit of the year.

L: Express the Mersenne prime $2^{32582657} - 1$ in base-eight notation, and use the first digit.

M: Find the penultimate digit of the trillionth Fibonacci number.

N: Find the minus-third Fibonacci number.

O: Add the two last digits of the Mersenne prime $2^{32582657} - 1$.

P: Find the last digit of $2 \uparrow \uparrow \uparrow \uparrow \uparrow 2$.

Q: There is a sequence starting with the DVD number, followed by the cult number with a film about it, then the ninth Fibonacci number, then the Ultimate Answer to Life, the Universe and Everything, then the difference between 1956 and 2006. Find the next number in this sequence, and use its first digit.

R: Find a number D between 1 and 32040 such that $17 \times D = 1$ (modulo 32040). Use its middle digit.

S: In the RSA system, with N = 143, M = 60, E = 11, find C.

T: Find the highest common factor of 12345679 and 888888; use its first digit.

U: Find the next number in the sequence beginning 19, 104, 22, 227, 129, 193, 241; use its last digit.

V: Find the 65537th digit after the decimal point of 1/12345679.

W: Find the last digit of $9^{7^{5^3}}$.

X: If the Time Traveller arrives on the first Friday of January 802701, and the Gregorian calendar is still in force, which day of the month is this? (January 1, 1900, was a Monday.)

Y: Find the next number in the sequence beginning 19, 104, 177, 22, 161, 232, 39; use the last digit.

Z: A two-digit Fibonacci number appears at the beginning of *Nineteen Eighty-Four.* Use the second digit.

Notes, links, updates and
answers to all problems are on
the Web at

www.cryptographic.co.uk/onetonine

Index of Main Subjects